The Frontlines of Artificial Intelligence Ethics

This foundational text examines the intersection of AI, psychology, and ethics, laying the groundwork for the importance of ethical considerations in the design and implementation of technologically supported education, decision support, and leadership training.

AI already affects our lives profoundly, in ways both mundane and sensational, obvious and opaque. Much academic and industrial effort has considered the implications of this AI revolution from technical and economic perspectives, but the more personal, humanistic impact of these changes has often been relegated to anecdotal evidence in service to a broader frame of reference. Offering a unique perspective on the emerging social relationships between people and AI agents and systems, Hampton and DeFalco present cutting-edge research from leading academics, professionals, and policy standards advocates on the psychological impact of the AI revolution. Structured into three parts, the book explores the history of data science, technology in education, and combatting machine learning bias, as well as future directions for the emerging field, bringing the research into the active consideration of those in positions of authority.

Exploring how AI can support expert, creative, and ethical decision making in both people and virtual human agents, this is essential reading for students, researchers, and professionals in AI, psychology, ethics, engineering education, and leadership, particularly military leadership.

Andrew J. Hampton is an Assistant Professor of Psychology at Christian Brothers University, US. He has served as Project Manager on the pioneering hybrid tutor ElectronixTutor, Chair of the IEEE Standards Association working group for Adaptive Instructional Systems, and Project Co-leader for a novel form of conversation-based AI learning and engagement called TalkShop.

Jeanine A. DeFalco is a Research Scientist with the US Army Futures Command, Combat Capability Center in Orlando, Florida, US. She serves as Chair of the IEEE Standards Association working group for Recommended Practice for Ethically Aligned Design of Artificial Intelligence (AI) in Adaptive Instructional Systems. Dr. DeFalco is also an Adjunct Instructor at the University of Central Florida, and Pace University, US.

The Frontlines of Artificial Intelligence Ethics

Human-Centric Perspectives on Technology's Advance

Edited by
Andrew J. Hampton, PhD and
Jeanine A. DeFalco, PhD

NEW YORK AND LONDON

Cover image: Getty

First published 2022
by Routledge
605 Third Avenue, New York, NY 10158

and by Routledge
4 Park Square, Milton Park, Abingdon, Oxon OX14 4RN

Routledge is an imprint of the Taylor & Francis Group, an informa business

© 2022 selection and editorial matter, Andrew J. Hampton and Jeanine A. DeFalco; individual chapters, the contributors

The right of Andrew J. Hampton and Jeanine A. DeFalco to be identified as the authors of the editorial material, and of the authors for their individual chapters, has been asserted in accordance with sections 77 and 78 of the Copyright, Designs and Patents Act 1988.

All rights reserved. No part of this book may be reprinted or reproduced or utilised in any form or by any electronic, mechanical, or other means, now known or hereafter invented, including photocopying and recording, or in any information storage or retrieval system, without permission in writing from the publishers.

Trademark notice: Product or corporate names may be trademarks or registered trademarks, and are used only for identification and explanation without intent to infringe.

British Library Cataloguing-in-Publication Data
A catalogue record for this book is available from the British Library

Library of Congress Cataloging-in-Publication Data
Names: Hampton, Andrew J., editor. | DeFalco, Jeanine A., editor.
Title: The frontlines of artificial intelligence ethics : human-centric perspectives on technology's advance / edited by Andrew J. Hampton and Jeanine A. DeFalco.
Description: Abingdon, Oxon ; New York, NY : Routledge, 2022. | Includes bibliographical references and index.
Identifiers: LCCN 2022005677 (print) | LCCN 2022005678 (ebook) | ISBN 9780367467661 (hardback) | ISBN 9780367467678 (paperback) | ISBN 9781003030928 (ebook)
Subjects: LCSH: Artificial intelligence—Moral and ethical aspects. | Artificial intelligence—Social aspects.
Classification: LCC Q334.7 .F76 2022 (print) | LCC Q334.7 (ebook) | DDC 174/.90063—dc23/eng20220421
LC record available at https://lccn.loc.gov/2022005677
LC ebook record available at https://lccn.loc.gov/2022005678

ISBN: 9780367467661 (hbk)
ISBN: 9780367467678 (pbk)
ISBN: 9781003030928 (ebk)

DOI: 10.4324/9781003030928

Typeset in Sabon
by codeMantra

Contents

List of Contributors   vii

Once More, with Feeling: Approaching a Novel Field of Technological Inquiry with the Hard-Earned Lessons of the Humanities   1
ANDREW J. HAMPTON AND JEANINE A. DEFALCO

PART 1
Surveying the AI Landscape   7

1 AI and the Crisis of the Self: Protecting Human Dignity as Status and Respectful Treatment   9
OZLEM ULGEN

2 *Flipping Rocks and Pointing out the Bugs:* Invisible Threats and Data   34
DARIAN J. DEFALCO

3 Seeing the Forest and the Trees: AI Bias, Political Relativity, and the Language of International Relations   45
DR. LEAH C. WINDSOR

PART 2
AI in the Classroom   61

4 Truth in Our Ideas Means the Power to Work: Implications of the Intermediary of Information Technology in the Classroom   63
COL JAMES NESS, LTC LOLITA BURRELL AND DAVID FREY

5 Benefits and Potential Issues for Intelligent Tutoring
  Systems and Pedagogical Agents   84
  LISHAN ZHANG, XIANGEN HU, FRANK ANDRASIK,
  AND SHUO FENG

6 The Only Living Boy in Homeroom: How Virtual
  Classes and Agents Fundamentally Change the
  Learning Experience   102
  ANDREW J. HAMPTON, DONALD "CHIP" MORRISON
  AND BRENT MORGAN

PART 3
Decisions, Decisions   123

7 J.A.R.V.I.S., What Should I Do Now? Human Virtual
  Agents as a Means to More Ethical Decision-Making   125
  JEANINE A. DEFALCO AND JOHN HART

8 Ethical Frameworks for Cybersecurity: Applications for
  Human and Artificial Agents   141
  F. JORDAN RICHARD SCHOENHERR AND ROBERT THOMSON

9 Deep Blue Wants You: Identifying and Addressing
  Sources of Bias in AI Systems to Support Human
  Resources Decisions   162
  ARYN PYKE, F. JORDAN RICHARD SCHOENHERR, AND
  ROBERT THOMSON

10 On the Shoulders of Giants: How the Social Sciences
  Can Help AI Navigate Its Ethical Dilemmas   185
  ARTHUR C. GRAESSER AND JOHN SABATINI

Index   203

Contributors

Frank Andrasik is a Distinguished Professor and former Chair (2010–2020) of Psychology and an Affiliate Faculty Member of the Institute for Intelligent Systems, University of Memphis, TN, US. His research has long focused on a number of domains within behavioral medicine and more recently on polytrauma in military populations, health disparities, and intelligent agents to facilitate learning. He belongs to a number of professional societies, holding the elected status of Fellow in seven of them: Society of Clinical Psychology and Society for Health Psychology, American Psychological Association; Association for Psychological Science; Society of Behavioral Medicine; Association for Behavioral and Cognitive Therapies (ABCT); American Headache Society; and Association for Applied Psychophysiology and Biofeedback (AAPB). In March 2019, he was presented a Lifetime Achievement Award by AAPB in recognition of his outstanding contributions to this society and the field of biofeedback. He has twice served as President for scientific societies: AAPB (1993–1994)and ABCT (2009–2010). Dr. Andrasik has published over 300 articles and chapters and 8 co-edited/co-authored texts; and presented numerous invited national and international addresses and workshops.

LTC Lolita Burrell began her military career at the Army Medical Department Officer Basic Course at Fort Sam Houston, Texas, where she was commissioned as a Captain in the Medical Service Corps. Prior to entering active duty, she completed her B.S. in Psychology at Delaware State University and Ph.D. in Medical Psychology at the Uniformed Services University of the Health Sciences. From 1999 to 2003, LTC Burrell served at The Walter Reed Army Institute of Research studying Soldier and Family health and readiness issues. From 2003 to 2007, she served as the Chief of the Medical Research Unit for The United States Army Research Institute of Environmental Medicine at their satellite facility at Fort Bragg. Next, she served as Faculty Member from 2007 to 2012 at The United States Army Military

Academy (USMA). LTC Burrell was a member of the Joint Mental Health Advisory Team-8 that deployed to Afghanistan prior to her assignment as the Deputy Director of the Plans, Programs, Analysis and Evaluation Directorate at The United States Army Medical Research and Materiel Command. Currently she serves as an Academy/Associate Professor at USMA. LTC Burrell is a member of the Army Acquisition Corps and has a certificate in Advanced Critical Incident Stress Management.

Darian J. DeFalco has been working in technology for over 20 years, including six and a half years at Cold Spring Harbor Laboratory, where he managed a team responsible for the computational and storage needs which support their mission as the world's leading institute for molecular biology. He is presently a Senior Site Reliability Engineer at Transfix, Inc., a startup which uses machine learning to optimize freight and logistics, both as a broker and as a service provider. He presently lives on Long Island with his wife and three children.

Jeanine A. DeFalco is a Research Psychologist (Adaptive Training) with the Army Futures Command, CCDC-STTC, Orlando, conducting research and supporting the development of the Army's Generalized Intelligent Framework for Tutoring (GIFT). Dr. DeFalco sits on the executive committee of the International Society for AI in Education (2020) and has been an active member of IEEE's working group to develop standards for adaptive instructional systems (AISs) (2018). She is currently the Chair for IEEE P2247.4, *Recommended Practice for Ethically Aligned Design of Artificial Intelligence (AI) in Adaptive Instructional Systems*. In 2020, Dr. DeFalco was awarded with her Co-presenter, Dr. Robert Sottilare, for best workshop on Adaptive Instructional Systems at the 2020 vI/ITSEC conference. In 2019, Dr. DeFalco was recognized with the NTSA Modeling and Simulation Award, Education/Human Performance—Team Award, for outstanding achievement in modeling and simulation for her contributions on The Generalized Intelligent Framework for Tutoring (GIFT). Dr. DeFalco received her Ph.D. in Psychology from Columbia University (2017), specializing in Human Development/Cognitive Studies in Education with a concentration in Intelligent Technologies. Dr. DeFalco also holds a Master's in Educational Theatre from New York University, a Master's in Drama Studies from Johns Hopkins University, and a Bachelor's in History and Theatre from Long Island University.

Shuo Feng is a Master Student in Education Technology at Central China Normal University. His research interests include intelligent tutoring systems and learning analytics.

David Frey is Professor of History and Founding Director of the award winning Center for Holocaust and Genocide Studies (CHGS) at the United States Military Academy at West Point. As Director of the CHGS, Dr. Frey has spearheaded efforts to increase Academy, Army and Defense Department awareness of, understanding of, research into, and efforts to prevent mass atrocity. He is the author of *Jews, Nazis, and the Cinema of Hungary: The Tragedy of Success, 1929–1944* (IB Tauris, 2017) and co-author of *Ordinary Soldiers: A Study in Law, Ethics and Leadership*. Dr. Frey earned his Ph.D. in Central European History from Columbia University. Prior to West Point, he taught at Columbia University. He serves on the US Holocaust Memorial Museum's Education Committee and is a Founding Executive Committee Member of the Consortium of Higher Education Centers of Holocaust, Genocide, and Human Rights Studies.

Arthur C. Graesser is Emeritus Professor in the Department of Psychology and the Institute of Intelligent Systems at the University of Memphis. He conducts research in discourse processing, cognitive science, educational psychology, computational linguistics, and artificial intelligence in education. He develops and tests software that integrates advances in these areas such as AutoTutor, a system that helps people learn with conversational agents.

Andrew J. Hampton is an Assistant Professor of Psychology in the Behavioral Sciences Department at Christian Brothers University. He has served as Project Manager on the pioneering hybrid tutoring system *ElectronixTutor*, Chair of the IEEE Standards Association working group for Adaptive Instructional Systems, Leadership Team Member for the NSF-funded Learner Data Institute, and Project Co-leader for a novel form of conversation-based AI learning and engagement called TalkShop. He has experience in classroom instruction for general psychology, writing for psychology, psychology of film, personality psychology, and advanced statistics. His research interests include technologically mediated communication, psycholinguistics, semiotics, adaptive educational technology, artificial intelligence, political psychology, and, of course, the ethical implications of AI from a psychological perspective.

John Hart is the Program Manager for the Army's University Affiliated Research Center at the University of Southern California specializing in the research, design, and development of simulations involving virtual humans for learning, education, and training. He holds a Doctor of Philosophy in Modeling and Simulation focusing on human experiences with virtual humans in social situations. Dr. Hart developed multiple virtual human experiences aimed at training skills for

Army leaders in social interaction. Some of the social skills training experiences include bilateral negotiations, leadership counseling, and discussions related to dealing with sexual harassment within a unit.

Xiangen Hu is a Professor in the Department of Psychology, Department of Electrical and Computer Engineering and Computer Science Department at The University of Memphis (UofM) and Senior Researcher at the Institute for Intelligent Systems (IIS) at the UofM and is Professor and Dean of the School of Psychology at Central China Normal University (CCNU). Dr. Hu received his M.S. in Applied Mathematics from Huazhong University of Science and Technology, and M.A. in Social Sciences and Ph.D. in Cognitive Sciences from the University of California, Irvine. Dr. Hu is the Director of Advanced Distributed Learning (ADL) Partnership Laboratory at the UofM, and is a Senior Researcher in the Chinese Ministry of Education's Key Laboratory of Adolescent Cyberpsychology and Behavior. Dr. Hu's primary research areas include Mathematical Psychology, Research Design and Statistics, and Cognitive Psychology. More specific research interests include General Processing Tree (GPT) models, categorical data analysis, knowledge representation, computerized tutoring, and advanced distributed learning. Dr. Hu has received funding for the above research from the US National Science Foundation (NSF), US Institute of Education Sciences (IES), ADL of the US Department of Defense (DoD), US Army Medical Research Acquisition Activity (USAMRAA), US Army Research Laboratories (ARL), US Office of Naval Research (ONR), UofM, and CCNU.

Brent Morgan, Ph.D., is the President and Co-founder of MagicStat.co and a Visiting Assistant Professor at Rhodes College. His primary research interest is in bi-directional adaptability in human–computer interaction. He has also investigated how emotions influence cognition across a variety of tasks. He has led or contributed to numerous research projects in artificial intelligence in education, including ElectronixTutor, the first hybrid intelligent tutoring system (ITS), and the Personal Assistant for Life-Long Learning (PAL3). He is currently developing a hybrid ITS to complement an undergraduate statistics course in Psychology.

Donald "Chip" Morrison is a lifelong Teacher and Educational Researcher. A graduate of Dartmouth College (Drama), he holds a Master's Degree in Language Studies from the University of Hong Kong, and a Doctorate in Education (Human Development) from the Harvard Graduate School of Education. Dr. Morrison began his career teaching English in Hong Kong, first at Hong Kong Baptist College and then at the University of Hong Kong. In the 1990s, while serving as a Senior Scientist at Bolt Beranek and Newman, he helped found Co-nect, a K-12 school reform network funded by New American Schools. This led to a specialization in methods of measuring school-level instructional quality, including school walkthroughs by peers. He has also completed stints as a Museum Exhibit Developer, and independent educational Software Developer. Dr. Morrison is recently retired from the Institute for Intelligent Systems, University of Memphis, where he held the title of Research Assistant Professor and conducted studies of human tutoring. His recent book, *The Coevolution of Language, Teaching, and Civil Discourse among Humans* (Palgrave MacMillan), discusses the close relationship between teaching and human language.

COL James Ness is currently assigned as an Academy Professor in the Department of Behavioral Sciences and Leadership at the US Military Academy, West Point. He earned the academic rank of Full Professor in 2016. COL Ness's prior assignment was as Command Inspector General, NATO Training Mission/Combined Security Transition Command – Afghanistan. COL Ness earned a Bronze Star for his efforts in reforming the Afghan National Military Hospital and establishing an internal assessment program within the Ministry of Interior. The latter informed Afghan leadership of the effectiveness of police force transition initiatives to Afghan control. COL Ness has had varied assignments in medical research and development of which his work resulted in a change to the ANSI and ICNRP safety standards for long-term viewing of near IR laser sources.

Aryn Pyke is an Associate Professor in the Engineering Psychology Program at West Point, and a Cognitive Scientist with the Army Cyber Institute. Dr. Pyke holds Bachelor's and Master's degrees in Electrical and Computer Engineering and a Ph.D. in Cognitive Science. Her research interests include STEM education and human–computer interaction. Her current focus is on the human-in-the-loop in cybersecurity and AI-human teaming contexts.

John Sabatini is a Distinguished Research Professor in the Institute for Intelligent Systems and the Department of Psychology at the University of Memphis. Dr. Sabatini's research interests are in reading literacy development and disabilities, assessment, cognitive psychology, the learning sciences, and educational technology. His research agenda spans developing and evaluating EdTech for reading, writing, and critical thinking in adolescents; aiding English language learners to develop reading and language skills; and understanding, instructing, and assessing learners of all ages with reading-difficulties. He provides technical and research advice to national and international surveys.

F. Jordan Richard Schoenherr is an Assistant Professor in the Department of Psychology, Concordia University, and an Adjunct Research Professor and a member of the Institute for Data Science, Carleton University. Dr. Schoenherr's primary areas of interest are learning and decision-making with application organizational behavior (ethical decision-making, incivility, insider threat), cybersecurity social dilemmas, and artificial intelligence (ethical AI and XAI). In his forthcoming book, *Trust in the Age of Entanglement* (Routledge Publishers), he provides a synthesis between the popular science and cognitive science of autonomous and intelligent systems. He has worked as a Visiting Scholar at the US Military Academy (West Point), a Senior Ethics Advisor at the Canadian Border Service Agency, Ethics Advisor for the Ombudsman, Integrity, and Resolution Office (Health Canada/Public Health Agency of Canada), and as the Primary Researcher for the scientific integrity initiative at the Office of the Chief Scientist (Health Canada).

Robert Thomson serves as the Cyber and Cognitive Science Fellow at the Army Cyber Institute and is an Associate Professor in Engineering Psychology in the Department of Behavioral Sciences and Leadership at the United States Military Academy. Dr. Thomson has over 9 years of post-graduate experience and over 40 invited and refereed academic publications and one patent in the domains of computational modeling, intelligence analysis, cybersecurity, and artificial intelligence. He has received funding from Intelligence Advanced Research Projects (IARPA), Defense Advanced Research Projects Agency (IARPA), Defense Advanced Research Projects Agency (DARPA), ONR, and ARL.

Ozlem Ulgen is Reader in International Law and Ethics, School of Law, Birmingham City University, UK. Dr. Ulgen specializes in moral and legal philosophy, weapons law, international humanitarian law, and public international law. She has published works on cosmopolitan ethics in warfare, Kantian ethics and human dignity in the age of artificial intelligence and robotics, and the law and ethics of autonomous weapons. She has a forthcoming publication with Routledge, *The Law and Ethics of Autonomous Weapons: A Cosmopolitan Perspective*. She is Principal Investigator for an EPSRC-funded project, "God, the Oracle, and the Nightclub Bouncer: Can Human Dignity be Modelled in an AI-based Decision Support System for post-Covid Health Certification?" She sits as an Expert on several international standard-setting and regulatory initiatives in the area of AI and robotics, including IEEE ECPAIS, the UN Group of Governmental Experts on Lethal Autonomous Weapons Systems, IEEE Standards Working Groups P7007 and P7000.

Dr. Leah C. Windsor is an Associate Professor in the Applied Linguistics group in the Department of English and the Institute for Intelligent Systems at The University of Memphis. She directs the languages across cultures and languages across modalities labs and is PI on an NSF grant studying multimodal signals in world leaders' speeches. Her interdisciplinary approach to understanding political language is situated at the intersection of political science, linguistics, and cognitive science. Her book on bias in family formation in academia, *The PhD Parenthood Trap: Caught Between Work and Family in Academia* (with Dr. Kerry Crawford), was published in October 2021 by Georgetown University Press. She is a 2020–2021 Non-Resident Fellow for the Krulak Center in the Marine Corps University.

Lishan Zhang is an Associate Professor at Central China Normal University. He received a Ph.D. in Computer Science from Arizona State University. He has published over 30 peer-reviewed academic papers. Dr. Zhang's research interests include intelligent tutoring systems, student modeling for personalized learning and educational data mining.

Once More, with Feeling

Approaching a Novel Field of Technological Inquiry with the Hard-Earned Lessons of the Humanities

Andrew J. Hampton and Jeanine A. DeFalco

1 Introduction

"We're just not addressing the human component." It was the kind of sweeping statement fitting for the setting of our conversation—a pub in North London—but generally lacking empirical support. We—Andrew Hampton and Jeanine DeFalco[1]—were halfway through a weeklong confederation of conferences dubbed the London Festival of Learning and we had become increasingly frustrated by the seemingly exclusive ethical focus on data privacy and ownership in learning systems. This trend persisted despite the pronounced presence of researchers in the humanities, notably psychologists like us. Yes, these are critical issues. Yes, they get considerably thornier when most users are minors (as tends to happen in the educational technology field). But certainly the advances in artificial intelligence (AI) being paraded around the festival engendered broader discussion than data anonymization and security. The ethical issues were everywhere—learning gains from incorporating AI; language patterns in human–AI interactions; automatically generated feedback to learner input; training metacognitive skills through intelligent recommendations—but people generally failed to ask what that meant outside of capability or efficiency. When those discussions did happen, they seemed to take the form of "We may *also* consider...". To our eyes, no one put human-centric ethics up front, or so much as cited an ethical framework.

Here our pub-based complaint session diverged from the usual pattern, because we decided to follow up. In surveying the field, we perceived a disconnect between the groups who should have been converging to tackle emergent questions. Social scientists and developers seemed to be cooperating no further than was necessary to create working, pedagogically sound systems. The often daunting technical challenges inexorably draw focus from the former group into the latter's bailiwick, and correspondingly away from the social component of social science. Meanwhile, ethical philosophers had no intentional presence.

DOI: 10.4324/9781003030928-1

But perhaps we were viewing the problem from the wrong end. Certainly AI ethics is and has been a prominent focus in the research community at-large. During the production of this volume, no less an authority than the Pope called for more work in the field (Copestake, 2020). But our reading only served to throw sharper focus on the lack of investigation into the psychological impact. Our survey indicated three primary areas of AI ethics research: (1) data privacy and ownership; (2) job displacement and economic changes; and (3) mortal threat (or what we like to call the "don't blow up the world" camp). Among the most prominent and widely cited writers in this field—boasting endorsements from Elon Musk and Bill Gates on his book cover—Bostrom (2014) summarized the field thusly:

> Yes, there is more interest in thinking about the consequences of advances in machine intelligence; but much of this ends up focusing on nearer-term concerns such as lethal autonomous weapons, labor market impacts of automation, cybercrime, privacy, or self-driving cars. These are not unreasonable things for some people to be thinking about, but they mostly concern issues quite distinct from those raised by human-level AI or superintelligence.

He then proceeds to fill part of this gap by cataloguing the various ways in which superintelligent AI (i.e., greater than human-level AI) could endanger the fate of our species. He does not, however, address the everyday impacts of these technologies on the people using them.

So we pushed forward. Despite the lack of published research with this focus, we found an appetite for the discussion. Our session proposal for the next year's International Convention of Psychological Sciences (ICPS), titled *Psychology, Ethics, and Artificial Intelligence*, was accepted without revision. It drew the largest crowd of either of our careers, even with a fully American panel presenting in Paris. From there, the momentum to create a book felt natural, and we began recruiting.

2 Assembling the Team

Plotting a cohesive narrative out of our perception of a literature gap proved a difficult task. As our ICPS session title indicates, our initial focus was limited to psychology. However, significant limitations of this approach quickly made themselves known, largely owing to the tremendously interdisciplinary (arguably transdisciplinary) nature of AI. It was not accidental that so many psychologists had found themselves at computer science conferences, but neither was the concern we articulated limited to psychological impact. One example came from the problem of providing equitable decision-support applications derived from machine learning. We had seen this discussed as a technical problem to resolve, but had not been formally exposed to the human cost of failure until we asked. This indicated broad legal ramifications. Similarly, the data necessary to drive these applications had a human component outside of our expertise.

To acknowledge the breadth of issues that had not been adequately addressed, we decided to begin with an introductory section drawing from disparate areas of the humanities. Chapters in the first section of this volume, *Surveying the AI Landscape*, do not neatly stay within department titles, consistent with the radical reimagining of disciplinary lines inherent in AI. Or, as a wise dissertation advisor quipped, science does not proceed according to whoever named the buildings at Oxford (V. L. Shalin, personal communication, many occasions). However, these chapters all comfortably reside beyond a strict definition of psychology. Contributors Ulgen, D. DeFalco, and Windsor raise issues and propose applicable moral frameworks within the fields of law, history, linguistics, and political science, providing necessary context for efforts in data science and machine learning.

With that brief tour of the human-centric AI in place, we turned toward our respective areas of AI expertise—conversation-based instruction (Hampton) and decision-support (DeFalco). We both felt (and feel) that these are broad, important, and pressing enough subfields to warrant sections within the book. Further, our extensive exposure to disparate approaches revealed the range of contradictory perspectives, even among close colleagues. We include these contradictions not as a bug, but a feature. The second section of this volume, *AI in the Classroom*, begins with Ness, Burrell, and Frey's historical philosophical review of the relative capabilities and distinctions between organic and AI, with prescriptions for how that can be best applied to learning writ large. This lays the groundwork for the two following chapters (by Zhang, Hu, Andrasik, & Feng and Hampton, Morrison, & Morgan, respectively), each of which discusses the mechanics of implementing conversational instruction through anthropomorphic agents. However, these chapters make divergent recommendations that present the reader with the task of critically evaluating their relative merits. We find this divergence particularly compelling because several of the authors are currently working together on multiple such systems.

The third section, *Decisions, Decisions*, takes three distinct approaches to how AI can support human decision making. This begins with an agent-based structure for making ethical decisions in a domain that frequently presents ethical dilemmas—the military. DeFalco and Hart detail the many challenges that must be overcome, but keep an eye on the potentially transformative influence of equipping every military service member with a knowable, accountable moral compass. Next,

Schoenherr and Thomson highlight the inescapably cognitive and social basis of cybersecurity. They point out the inadequacy of the presently dominant analogical paradigm for understanding threats, particularly when more adaptable, accurate, and actionable schemata are available. The section concludes with Pyke, Schoenherr, and Thomson delving deeply into the problem of providing equitable decision support derived from machine learning, and the substantial impact this has in an array of domains. They conclude that successful remediation will require improvements on both the human and AI sides.

Finally, a concluding chapter from eminent researchers Art Graesser and John Sabitini highlights threads running throughout the book to argue for frequent and explicit reference to precedents established by social science when tackling novel issues. We note here the variation in vocabulary between their chapter and this one, i.e., humanities versus social sciences. This disparity is intentional and reflects: (1) the inherently overlapping and indistinct boundary between these terms (see Homans, 1961); (2) a slight divergence of emphasis, essentially between a desire for broad perspective and a dictate for empirical rigor, respectively; and (3) an embrace of disagreement inherent in investigating a novel domain. Graesser and Sabatini organize their reflections around several driving questions that will require periodic reevaluation based on the advancement and adaptation of AI technologies. Based on that dynamic relationship, they make the case that the best path forward is not prescriptive, but rather a framework of complementary human–AI evaluation by which we can make informed decisions regarding necessary tradeoffs.

3 Conclusion

AI is already effecting our lives profoundly, in ways great and small, obvious and opaque. Much academic and industrial effort has considered the implications of this AI revolution from technical and societal perspectives, but we find that the humanistic impact of these changes has primarily been relegated to anecdotal evidence in service to a broader frame of reference. Explicit examination of how people interact with AI—through the lens of the humanities—constitutes a vital conversation. Currently, these conversations are largely fractured and confined to those development teams lucky enough to have a stray member with a liberal arts degree. These discussions must be brought to wider attention, and presented in a way that engages stakeholders and decision makers beyond academia, beyond computer scientists, and into the active consideration of those in positions of authority with respect to policy decisions. The contributions in this book attempt to begin that process. With feeling.

Note

1 Andrew's wife Kathryn was also present at this initial discussion, and made critical contributions in pointing out when we had drifted off topic or ceased to make sense to someone outside our research niche. Her medical expertise (as a doctor of nursing practice) also provided a practitioner's perspective in a domain well-charted by classical ethics.

References

Bostrom, N. (2014). *Superintelligence: Paths, dangers, strategies.* Oxford University Press.

Copestake, J. (2020, 28 February). AI ethics backed by Pope and tech giants in new plan. *BBC News.* https://www.bbc.com/news/technology-51673296.

Homans, G. C. (1961). The humanities and the social sciences. *American Behavioral Scientist,* 4(8), 3–6.

Part I
Surveying the AI Landscape

Chapter 1

AI and the Crisis of the Self
Protecting Human Dignity as Status and Respectful Treatment

Ozlem Ulgen

> ... a great deal of the work which was formerly done by human beings is now being done by machinery. This machinery belongs to a few people: it is being worked for the benefit of those few, just the same as were the human beings it displaced. These Few have no longer any need of the services of so many human workers, so they propose to exterminate them! The unnecessary human beings are to be allowed to starve to death! And they are also to be taught that it is wrong to marry and breed children, because the Sacred Few do not require so many people to work for them as before!
> (Tressell, 2004, p. 114)

1 Introduction

Over a century since Robert Tressell's prescient novel, the unsettling reality of technology replacing humans continues. A tidal wave of messianic worship for AI, robotics, "Big Data," "Internet of Things" is upon us, mainly articulated through the efficiency paradigm—improving productivity, enhancing human capabilities, reducing time spent on mundane tasks. From algorithms that determine student grades, personalize online marketing, approve financial credit applications, assess pre-trial bail risk, and select human targets in warfare, it seems we are willingly complicit in relinquishing decision-making powers to machines. As Tressell reminds us, we need to understand who "These Few" are controlling the technology and to what purpose it is put, rather than completely repudiate technological innovation (2004). The Nobel Prize-winning economist Joseph Stiglitz (2018) warns that without governmental policies that support sharing of increased productivity from AI across society, there will be rising unemployment, lower wages, and acute social inequalities. Against this backdrop of political, social, and economic challenges, viewed from a moral philosophical perspective, unfettered use of AI that diminishes human agency and decision-making powers undermines human dignity. Detrimental impact of AI on human dignity

DOI: 10.4324/9781003030928-3

is not so easily understood, especially when its justification is presented as some sort of a gain for humanity; saving time, energy, or delegating routine tasks. But human interaction that is mediated by technology penetrates the core of what it means to be human; autonomy and agency to engage in free-thinking, and exercise reasoning, judgement, and choice. This is the moral value of human dignity.

In this chapter, I argue that human dignity is a universal moral value that should be at the center of policy formulation and laws governing AI innovation and impact on societies. Part 2 sets out concerns about AI innovation and its potential adverse impact on human dignity. Part 3 considers how diverse cultures, international legal instruments, and constitutional laws represent human dignity as innate human worthiness that is a universal moral value, a right, and a duty. Part 4 develops two distinct dimensions of human dignity which can be concretized in policy and law relating to AI: (1) recognition of the status of human beings as agents with autonomy and rational capacity to exercise reasoning, judgement, and choice; and (2) respectful treatment of human agents so that their autonomy and rational capacity are not diminished or lost through interaction with or use of the technology.

2 AI Innovation and Impact on Human Dignity

It is impressive how AI is being developed for use in different domains and real-life settings—algorithms determining student grades, personalizing online marketing, approving financial credit applications, assessing pre-trial bail risk, and selecting human targets in warfare. But is it morally right to be deploying AI in such scenarios when inanimate deterministic activities have human consequences? In the UK and Europe, the ongoing COVID-19 pandemic has meant students were unable to sit exams necessary for entry into university. Instead, predictive algorithms, relying on past student performance and averaging determined grades, led to anomalies, bias, and unfair results (Zimmermann, 2021). With clear consequences for future educational and employment prospects, it seems immoral and reckless to have algorithms performing grading functions that reduce individual students to mere statistics without applying human judgement. Applying data processing and personal data rights contained under the EU General Data Protection Regulation (GDPR, European Parliament and Council of the European Union, 2016), the Norwegian Data Protection Authority claimed the International Baccalaureate Organisation breached Articles 5(1)(a) and 5(1)(d) in using a profiling algorithm which did not process student grades fairly, accurately, and transparently (2020). It requested rectification of grades.

Pre-trial bail risk algorithms used to assist human decision-making may seem like good examples of human-machine interaction. But poor dataset reliance and automation bias on the part of the human result in unfair outcomes. In the United States, a pre-trial bail risk assessment algorithm—used by judges to decide whether to release a defendant on bail or to remand them in custody—has come under increasing scrutiny. Among others, the Pretrial Justice Institute, a nonprofit organization previously advocating use of algorithms instead of cash bail, withdrew support for their use because such algorithms perpetuate racial inequities (2020; Open Letter by Academics, 2019). And at the extreme end of warfare, an algorithm may be determining who should be selected and attacked as a military objective leading to injury and death (Ulgen, 2019b). Unfairness, inequalities, restrictions on liberty, and life or death decisions form a concerning list of real human consequences as a result of AI systems.

Reflecting on the relationship between man and technology, throughout human history societal changes occurred as a result of new knowledge and technological innovation. Economic historians refer to four phases of innovation shaping economic development: the mechanization of textile manufacturing; railroads and steam from 1840 to 1890; steel, engineering, and electricity from 1890 to 1930; and automobile, fossil fuel, and aviation from 1930 to 1990 (Freeman & Louçã, 2001; Rosenberg & Birdzell, 1986). AI-based technologies fall into the post-1990 economic development phase. This "fourth revolution" includes information and communication technologies, AI, and autonomous robotics impacting every aspect of our lives today (Floridi, 2014). Yet a single invention cannot be the sum of our lives, problems, or solutions.

The drive toward greater efficiency and increased productivity precipitates the AI innovation Ferris-wheel; a never-ending cycle of innovation to counter human fallibility that rewards slavish adoption and punishes the reticent human mind. Byung-Chul Han (2017) refers to this as "psychopolitics"; a form of control of the human psyche exerted by technological domination and use of personal data in the public and private spheres that alters our minds and behavior to an extent that undermines our autonomy and agency. If we are constantly having to sync different platforms, update new software, connect systems with systems so that we can access even bigger systems, we are losing sight of ourselves and getting entangled in a techno-bureaucracy purposely constructed by two strange bedfellows: the regulators and the hackers. Both contribute to the crisis of the self.

2.1 The Techno-Bureaucracy of Hackers and Regulators

Hackers want to explore and exploit new technology vulnerabilities to serve their own illicit purposes, thereby increasing demand for higher security measures from regulators. Regulators (seemingly concerned with human well-being and protection of rights) introduce layers of complexity through overlapping and competing non-legally binding and legally-binding rules, ethical principles, and processes contained in global, regional, and national ethical frameworks, standards, and instruments (e.g., GDPR, 2016; EU AI Guidelines, 2019; G20, 2019; IEEE, 2019; OECD, 2019; UN Secretary General's High Level Panel on Digital Cooperation, 2019; AI Act Proposal, European Parliament and Council of the European Union, 2021). Meanwhile, private sector corporate entities, the military, and the state continue to develop AI under the radar of any enforceable regulation.

It is unclear how divergent ethical/legal initiatives apply across jurisdictions and alongside national legislation. The rules, principles, and processes are often impenetrable to the ordinary person. Take for example the legal concept of "responsibility" determining who or what will be held liable for any harm/damage caused by the technology, AI has potential to disrupt the attribution and causation chains unless there is always a human who will be held responsible throughout AI design, development, and deployment stages.

Self-learning algorithms and robots present the spectre of harmful and unattributable behaviors, which at the same time undermine human agency of foresight, prudence, and judgement in taking action with consequences in mind. Although responsibility is a priority ethical value and legal requirement contained in several global, regional, and national regulatory frameworks, its interpretation and implementation differs.

The UK recognizes legal responsibility, accountability, and legal liability as key issues in application of the law to AI, but focuses on developing principles of accountability and intelligibility (which are not the same as legal responsibility or liability) with possible review of the adequacy of existing legislation on legal liability (UK House of Lords Select Committee, 2018). For China, although responsibility is a core principle applicable at both the AI development and deployment stages, it is situated within an ethical framework biased toward commercial exploitation for the purpose of domestic economic growth.

It is unclear who or what will be held legally responsible, and future policies/laws may contain a commercial intellectual property/trade secrets exemption preventing disclosure of algorithmic models, datasets, and algorithmic reasoning (Standards Administration of China, 2018).

2.2 Freeing or Enslaving?

Whether AI-based solutions to everyday tasks are freeing or enslaving impacts on the crisis of the self. Does AI free up the human mind to undertake qualitative judgement-based complex tasks instead of routine memorizing numbers, memory recall, and mental arithmetic? Or, is more time spent frustrated by the technology (how it works, errors it produces, and rectification of errors and seeking redress)? In theory, more AI-assisting jobs should be available leaving routine tasks to machines. In practice, such jobs are few and far between with not enough training offered by employers to make the transition from displacement by machine to human-machine teaming (e.g., Semuels, 2020).

Among other mental tasks, recall and mental arithmetic stimulate the brain. Arguably, if we become dependent on technology for the simplest of tasks, we are enslaved by the technology and forget how to function. Automation bias is a manifestation of such enslavement whereby in human-machine tasks, the human operator favors the machine's response over their own judgement with major repercussions for lives and livelihoods (Cummings, 2004; Raja & Dietrich, 2010).

De-skilling may also occur through automata behavior exhibited in humans reduced to binary responses without independent critical thinking or judgement. Studies show that heavy use of digital technologies cause neurological changes that impede comprehension, retention, and deeper thinking (DeStefano & LeFevre, 2007; Small & Vorgan, 2008; Sweller, 1999; Zhu, 1999). This diminishes human agency and dignity with potentially serious repercussions for other humans. Remote pilots of unmanned armed aerial vehicles, for instance, thousands of miles away from conflict zones viewing video images of targets to select and attack, have been shown to exhibit moral disengagement and lack of deeper thinking. They are less fearful of being killed and less inhibited to kill. They have problems identifying targets, and reduced situational awareness in complex scenarios resulting in civilian fatalities (Linebaugh, 2013; Power, 2013; Royakkers & van Est, 2010; Woods, 2015).

As for being frustrated by the technology, accessing your online bank account in 2020 can be an exercise fraught with technical and security glitches. You need at least two different devices: one to receive a verification code, another to enter the code in order to gain access. Accessing other online accounts for home, work, or personal purposes requires memorizing codes, online storage of codes and passwords, or using facial/voice/fingerprint recognition technology. The technology-based solutions have flaws such as high error rates, non-recognition of dialects and accents, bias, and security breaches of stored biometric data (Buolamwini & Gebru, 2018; Errattahia, Hannania & Ouahmanewe, 2018; Singer & Metz, 2019; Taylor, 2019). Personal data divulged and stored across different platforms and devices actually leads to a loss of control over what is happening.

The crisis of the self will continue unless we confront issues of control and use of AI, and determine what supports rather than undermines human dignity. Let us now consider how diverse cultures, international legal instruments, and constitutional laws represent human dignity as innate human worthiness that is a universal moral value, a right, and a duty.

3 Human Dignity as a Universal Moral Value, Right, and Duty

There is a long history of philosophical, religious, and legal thinking on human dignity; what it entails and how it manifests. Such thinking reflects on what it means to be human and shows a sensibility toward articulating human worthiness.

The ancient Romans used the concept of *dignitas* to differentiate persons of rank and elevated social status from the common people (Iglesias, 2001). In Christian theology, human dignity developed from the idea that human beings are created in the image of God and therefore possess worthiness and deserve to be treated with reverence (English Standard Version Bible, 2001, Genesis 1: 26–27;[1] Genesis 5:1;[2] Genesis 9:6[3]). Augustine of Hippo (354–430 CE), a North African bishop influential to early Christian thinking, considered it important to nurture and value the inner self in order to enable moral rules to emerge. In the 13th century, St. Thomas Aquinas (as per Project Gutenburg, 2006) identified rational nature as an intrinsic human quality which leads to personhood and dignity.

In Hinduism, human dignity is conceptualized as individual for all living things, and not just humans, albeit with different approaches as to how it is attained, ranging from a non-inclusive, class-based conception of human dignity to Gandhi's conception of the equality and dignity of all humans (Braarvig, 2014; Gandhi, 1948). In Confucianism human dignity functions as ethical conduct and relates to three qualities: benevolence, righteousness, and integrity (An'Xian, 2014).

Islam recognizes human dignity as a status bestowed by God on pious individuals who fulfil their obligations toward God. Human dignity means security and safety in the life of society, sanctity of life, and honor in the conduct of one's public and private life (Maróth, 2014). The African tradition of *ubuntu* is a communitarian-based notion of human dignity relating to social honor, group moral standing, and the capacity to form communal relations (Metz, 2014).

But the most sophisticated and secular notion of human dignity comes from the 18th-century deontological philosopher, Immanuel Kant.

3.1 Kantian Human Dignity

Kant makes an explicit connection between human existence and human dignity. His categorical imperative urges us to "act in such a way that you always treat humanity, whether in your own person or in the person of any other, never simply as a means, but always at the same time as an end" (Kant, 1785/1969, p. 91).

For Kant, human dignity is a special status conferred on humans by virtue of their innate worthiness as sentient beings with the capacity to engage in rational thinking to create and abide by rules. From this special status flows certain rights and duties toward development of one's own free will, fettered to avoid gratuitous encroachment on others' free will.[4] A point of objection here is that human dignity appears to exclude those who lack rational thinking capacity (e.g., children, mentally disabled, wrongdoers, criminals, and the deceased). But Kant's formulation is not intended to create an elite human class or exclude the vulnerable; rather, it rationalizes a secular human-centric approach that distills core elements of humanity that are capable of universalization (Ulgen, 2017). Thus, it is the *capacity* for rational conduct rather than actual rational conduct that entitles all to human dignity. The capacity of others to act rationally to create and abide by rules that protect the vulnerable is a manifestation of human dignity, and Kant provides specific rules regarding the treatment of wrongdoers, criminals, and the deceased (Kant, 1797/1996).

A conception of human dignity based on human innate worthiness and rational capacity affords universal application and grounding to recognize it as a universal moral value. Innate worthiness and capacity are not dependent on societal, national, state hierarchical structures to confer status in order to set rules governing human exchange and interaction. In recognizing the intrinsic worth of humans, Kantian human dignity does not require formal recognition of personhood by any institutional structure, and discounts arbitrarily determined extrinsic considerations of nationality, religion, wealth, gender, birthplace, or family connections. Neither wrongdoing nor criminality denies a person's human dignity. Intrinsic worth pre-exists in all humans and is the basis for their special status with rights and duties attached.

Some such as Waldron (2009) criticize Kant's emphasis on respect for the innate rather than the person. But Kant situates human worthiness in something innate in order to avoid contested notions of formalized personhood dependent on extrinsic recognition (e.g., by the state, or a community) and which may exclude certain categories of persons. Recognition of innate worthiness then leads to autonomy, and rational capacity to exercise reasoning, judgement, and choice.

3.2 Human Dignity in International Legal Instruments and Constitutions

References to innate worthiness, human value, and rational capacity are contained in international legal instruments and state constitutions which recognize human dignity as a universal moral value, a right, and a duty. The Universal Declaration of Human Rights (UDHR, UN General Assembly, 1948) provides an understanding of human dignity based on the Kantian notion; that it is intrinsic to all humans endowed with reason and conscience, and recognizable in humanity as a whole. The Preamble states, "Whereas recognition of the inherent dignity and of the equal and inalienable rights of all members of the human family is the foundation of freedom, justice and peace in the world," and "the dignity and worth of the human person," also repeated in the Preamble to the International Covenant on Civil and Political Rights (ICCPR, UN Treaty Series, 1966a), and the International Covenant on Economic, Social and Cultural Rights (ICESCR, UN Treaty Series, 1966b). Article 1 of the UDHR states, "All human beings are born free and equal in dignity and rights. They are endowed with reason and conscience and should act toward one another in a spirit of brotherhood." Article 22 protects a person's "economic, social and cultural rights indispensable for his dignity and the free development of his personality." Article 23(3) protects the right to just and favorable remuneration for work in order to ensure "an existence worthy of human dignity."

These international legal instruments show that human dignity is a pre-existing status of all humans by virtue of their innate worthiness, providing a rationale for protection of human rights. Human dignity also operates as a guiding principle for interpreting and applying rights. The rationale and guiding principle aspects can be seen in the ICCPR and ICESCR. Article 10 of ICCPR requires that "All persons deprived of their liberty shall be treated with humanity and with respect for the inherent dignity of the human person." The right to education, contained in Article 13 of the ICESCR, is necessary for "the full development of the human personality and the sense of its dignity."

Beyond the universal moral value, several states recognize human dignity as a right and a duty. It is represented as a pervasive norm as well as a duty under German constitutional law. Article 1(1) of the German Basic Law provides that "Human dignity shall be inviolable. To respect and protect it shall be the duty of all state authority." Article 79(3) prohibits any amendment to human dignity as a state duty (Federal Republic of Germany, 1949). The state has both a negative and a positive obligation. "Respect" requires the state to refrain from acts that violate human dignity, and it must "protect" individuals "against humiliation, branding, persecution, outlawing" from third party acts (Bundesverfassungsgericht, 2000). During proceedings of the Parliamentary Council that debated the content of constitutional provisions, Theodor Heuss referred to Article 1(1) as a "non-interpreted thesis" that was an important value yet open to different interpretations (1993). But against the backdrop of acts of dehumanization experienced under the Nazi regime, the drafters had a clear sense that individual humans needed to be at the center of state legislation and protection. By placing human dignity in the first article of the constitution and before exposition of fundamental rights, the drafters achieved this objective. It means human dignity is woven into the fabric of legislative interpretation and state structures. It requires constant reference and application to give it substantive meaning and effect in practice.

Aside from representing an overriding constitutional norm and one that is fundamental to protecting rights, human dignity has increasingly been interpreted as a standalone substantive right that guarantees a "dignified minimum existence." It encompasses both the physical existence of a human being as well as the possibility to maintain interpersonal relationships and a minimal degree of participation in social, cultural, and political life.[5] The Constitutional Court recognizes the right to human dignity means all human beings possess this dignity as persons, irrespective of their qualities, their physical or mental state, their achievements and their social status, or any wrongdoing.[6]

Human dignity as a fundamental value and right has been central to the development of South African constitutional jurisprudence. Section 1 of the 1996 South African Constitution provides that the Republic of South Africa is one, sovereign, democratic state founded on "Human dignity, the achievement of equality and the advancement of human rights and freedoms." It is affirmed as a "democratic value" of the Bill of Rights, and specifically identified as a right under Section 10 which states, "Everyone has inherent dignity and the right to have their dignity respected and protected." The South African Constitutional Court (2004, paragraph 41) has declared that "dignity is not only a value fundamental to our constitution, it is a justiciable and enforceable right that must be respected and protected." The Court has held that human dignity inherently includes protection of the family (Constitutional Court of South Africa, 2000); requires protection of the social and economic conditions of vulnerable populations so that state-funded educational benefits extend to certain non-citizens (Constitutional Court of South Africa, 2004); and requires the state to provide substantial resources in order to realize the right to adequate housing (Constitutional Court of South Africa, 2001).

Clearly, human dignity is a universal moral value and, in some jurisdictions, is also understood as a right and a duty. This provides justification for placing human dignity at the center of policy formulation and laws governing AI technologies and innovation. To understand how this can be achieved, let us now turn to developing the content of human dignity as a status and as respectful treatment.

4 Human Dignity as a Status and Respectful Treatment

Kant's secular theory provides the basis for developing two distinct dimensions of human dignity which can be concretized in policy and law relating to AI: (1) recognition of the status of human beings as agents with autonomy and rational capacity to exercise reasoning, judgement, and choice; and (2) respectful treatment of human agents so that their autonomy and rational capacity are not diminished or lost through interaction with or use of the technology.

4.1 Recognition of the Status of Human Beings

4.1.1 Agents with Autonomy

Recognizing that human agents have autonomy relates to how they perceive situations, their ability to take independent action, and to exercise choice. Maintaining human autonomy is clearly of concern for policy formulation and law governing AI.

Autonomy as a philosophical concept refers to the capacity for self-government to make decisions and take action. Individual rights represent autonomy as individual freedoms to take action. But autonomy does not operate in isolation from rules or others and therefore needs to be contextualized within a moral framework.

Kant's notion of autonomy, which he refers to as "autonomy of the will" (1785/1969, pp. 94, 101), involves individual freedom to self-govern by taking morally-informed decisions and actions. Human agents act autonomously to provide reason for taking action, and decipher what is moral and what is not. It is this internal capacity for morally-informed conduct, rather than sanctions imposed by the state, which leads to freedom and inculcates a sense of duty to act morally. This has implications for individuals formulating policy/law, as well as individuals using or interacting with AI.

A "technology-biased approach" (Ulgen, 2020) to regulation—focusing on AI capabilities and limitations to improve performance, optimize operational efficiency, and identify and rectify any errors or failures—will prove inadequate to recognizing human autonomy. Human wants, needs, and values should be incorporated into the AI system design, development, and deployment to maintain and recognize human autonomy. Moreover, the AI system should not erode the human internal capacity for morally-informed conduct by imposing technology-only solutions, or altering thinking and behavior to induce immoral conduct. The EU AI Guidelines (2019) provide an example of how autonomy can be protected by requiring that humans should be able to "keep full and effective self-determination over themselves, and be able to partake in the democratic process."

Another way to protect human autonomy is to allow for non-deterministic influences on decision-making such as environment, learning, and critical thinking. Prior to deployment and regulation of

AI, rationale needs to be provided for context and appropriateness of use taking account of non-deterministic influences. It is at this stage that policy-makers, legislators, and regulators have autonomy to decide what is legally and morally acceptable, and therefore bear responsibility for their actions. AI designers and developers would need to take a "human-centric approach" (Ulgen, 2020) to think about user awareness, rights, and to represent non-deterministic influences on decision-making. If the latter is not possible, users should be informed of the deterministic decision-making beforehand and given the option for human decision-making.

The potential human user's autonomy is protected if they are able to act influenced by reason; if they can identify the motivations prompting their action; or they can change their motivations if they cannot identify with them. If the human takes action not based on reason, cannot identify the motivations prompting their action, or cannot change their motivations, then these would indicate that autonomy has been lost. Automation bias and human automata behavior induced by the AI system are examples of how this can occur. Thus, an AI system that relies on binary conditions being met without consideration of context, personal circumstances, or judgement ends up undermining autonomy.

4.1.2 Agents with Rational Capacity to Exercise Reasoning, Judgement, and Choice

Rational capacity is a human characteristic that manifests in the ability to exercise reasoning, judgement, and choice. An infinite number of scenarios, human characteristics, circumstantial evidence, environmental factors and combinations of these influence whether and how a person acts and whether the act is moral. Human perception and social interaction enables deciding whether the act is moral, and requires applying rules or principles to that particular situation, not simply as a calculative or performative process but as part of reflective thinking.

To exercise reasoning is to draw conclusions from a set of premises. It is a dynamic, ongoing process that may rely on common-sense presumptions as well as synthetic *a priori* judgements (Kant, 1781/1998) where the predicate is external to the subject and adds something new to our conception of it.

Reasoning involves practical reasoning (i.e., what to do), and pure reasoning (i.e., pondering in abstract form). To cope with the infinite number of influencers on moral conduct, both types of reasoning are necessary. This is apparent in Kant's conception of reasoning as requiring universality (that a rule guiding moral conduct into action must be capable of being used by others, or universalized), internal capacity for morally-informed conduct, and an ability to engage in deliberative and reflective thinking (Kant, 1785/1969, p. 84; 1797/1996, p. 157).[7] Though machines and algorithms can engage in practical reasoning and work well with pre-programmed premises and assumptions, they cannot engage in pure reasoning that is whimsical, inconclusive, and resurfacing at a later stage to enable decision-making.

The tendency to revert to common-sense presumptions in practical reasoning is a shortcoming of both humans and machines programmed by humans (e.g., "AI is a new innovation that will lead to new challenges"), which Kant criticized as an "emergency help" "when one knows of nothing clever to advance in one's defense" (1783/2004, p. 9). On the other hand, engaging in critical and reflective thinking as to what these AI challenges might be and how to mitigate them is to engage in pure reasoning in order to reach synthetic *a priori* judgements. It is this pure reasoning capacity of humans that should be protected and ring-fenced from AI intrusion.

Judgement is the faculty of thinking, the particular is contained under a universal rule, principle, or law, and functions as an "intermediary between understanding and reason" (Kant, 1790/2000, pp. 64, 66). Kant identifies two types of judgement: "determinant" (where the universal rule, principle, or law is already known and the particular is easily subsumed under it); and "reflective" (where the particular is known but the universal rule, principle, or law has to be found for it). An example of determinant judgement is knowing that physically assaulting someone is morally wrong and unlawful. But it is reflective judgement that differentiates humans from machines. The value and purpose of human reflective judgement can be illustrated in the following example.

A state official's sworn affidavit states that prison conditions must satisfy a list of legal and ethical requirements to protect prisoners' well-being and, therefore, if the defendant were to be imprisoned, they would not be subject to inhumane, degrading, or life-threatening treatment. The existence of the list and its application to prisons are facts. However, it does not follow that the defendant, if imprisoned, would not suffer ill-treatment. There is a difference between what is stated on the list and what is actually implemented in practice.

Without evidence of how the legal and ethical requirements are implemented in prisons generally, and in the particular prison relevant to the defendant, and the effects on prisoners, the state official is in no position to make the determination that the particular defendant would not suffer ill-treatment. In fact, the affidavit is worthless, as the list of legal and ethical requirements could simply be reeled off by an algorithm as factors that a judge should take into consideration during sentencing. The point is that the human (state official or judge) is required to go beyond determinant judgement and engage in reflective judgement to consider whether the defendant would suffer ill-treatment. This is something that cannot be performed by algorithms. There are no pre-programmable or pre-existing universal rules that can be relied upon. Through reconciliation and calibration of understanding and reasoning, the human decision-maker is able to reach a judgement.

Others point to more specific features of human judgement that cannot be replicated in machines or algorithms. Suchman intimates consciousness as a requisite to exercising judgement when she refers to it as self-direction that cannot be specified in a rule (Suchman, 1985). Weizenbaum (1976) refers to it as wisdom which only human beings possess because they have to "confront genuine human problems in human terms." Judgement is often required in the grey areas, the problematic issue points where there is no precedent to follow and no clear-cut answer or solution. This implies a discretionary aspect devoid of orderly rule formation and adherence. But it also captures an invaluable human faculty leading to human solutions that reinforce human dignity in the person exercising judgement and the person affected by it.

For example, in a situation where there is an automated decision-making system deciding on an applicant's eligibility for a health test, the system may automatically reject the applicant because it detects incomplete or unclear information. A human decision-maker can exercise judgement to determine the significance of any incomplete or unclear information and therefore decide on an appropriate response which may not involve outright rejection of the application.

Finally, the exercise of choice, as a manifestation of human rational capacity, is the process by which different desires, pressures, and attitudes compete leading to a decision and action. Kant (1781/1998, p. 533; 1785/1969, p. 101) referred to choice as a competing process that is controlled by the self for higher purpose such as reason or morality. In regulating the use of AI, the EU refers to "meaningful opportunity for human choice" (EU AI Guidelines, p. 12). Thus, if a person does not want to use AI for resources or services, alternatives must be provided. Equally, if prior to use or during use of an AI system a person decides that they no longer want to be subject to automated decision-making, they must be allowed the opportunity to opt out or withdraw consent. This reflects enforceable rights under the GDPR and the Council of Europe's Modernized Convention for the Protection of Individuals with regard to Automatic Processing of Personal Data (Convention 108+; Ulgen, 2020).

4.2 Respectful Treatment of Human Agents

Through use of and interaction with technology, humans should not be subjected to forms of treatment which would undermine their human dignity or diminish their autonomy and rational capacity. This also relates to Kant's core notion of human dignity as not treating humans as mere means to ends. It follows that respectful treatment entails recognition of human agents' autonomy and rational capacity. This can manifest in the AI system in several ways: (a) respecting human agent rights; (b) respecting AI limitations; (c) respecting prioritization of human needs.

4.2.1 Respecting Human Agent Rights

For an AI system to respect human agent rights requires designers and developers of such systems to adopt a "human-centric approach" taking account of rights to privacy, data protection, and fundamental rights.

The privacy and data protection rights provided for under the GDPR and Convention 108+ are an obvious starting point due to their reach across the entire lifecycle of a system. These are rights to: not being subjected to automated decision-making; prior consent; prior notification of right to withdraw consent; notification of automated decision-making; access to personal data; access to information on the logic of an automated decision; information on the significance and envisaged consequences of automated decision-making; object to processing of data; lawful, fair, and transparent processing of data; rectification of inaccurate data; withdraw consent; explanation of automated decision; obtain human intervention; express a point of view; and contest an automated decision (Ulgen, 2020).

Among the fundamental rights, most relevant for an AI system to respect are: freedom from torture, cruel, inhuman or degrading treatment;[8] freedom of expression;[9] freedom of thought, conscience, and religion;[10] and freedom of association.[11]

It has been argued that the use of autonomous weapons in warfare is contrary to human dignity and constitutes a form of cruel, inhuman or degrading treatment because such weapons treat humans as disposable inanimate objects rather than beings with intrinsic value and rational capacity (Ulgen, 2019a).

Autonomous weapons are characterized by their use of AI and robotics in order to achieve varying degrees of autonomy in the critical functions of acquiring, tracking, selecting, and attacking targets. Human involvement, either partially or fully, may be preempted or removed in any of these critical functions, and from the lethal force decision-making process. Replacing human combatants with AI and robotics means human moral and legal agency is lost.

A hierarchy of human dignity is created whereby certain humans are deemed more valuable than others. The human combatant is protected from harm and their human dignity is elevated above that of the human target. The human target is treated as an inanimate object without any interests; easily removed and destroyed by a faceless and emotionless machine. All individuals targeted and killed by such weapons are entitled to respect for their human dignity. Whether or not they are designated enemy combatants or terrorists, they have rational capacity, possess a moral value of dignity which cannot be replaced by an equivalent, and they cannot lose such status through immoral acts.

Common Article 3 of the 1949 Geneva Conventions provides fundamental guarantees (applicable to both non-international and international armed conflicts) that civilians and *hors de combat* "shall in all circumstances be treated humanely" (International Court of Justice, 1986).[12] Enemy combatants are protected under Articles 1(2) and 75 of the 1977 Additional Protocol I to the Geneva Conventions, which refer to "the principles of international law derived from established custom, from the principles of humanity and from the dictates of public conscience"; and if they do not benefit from more favorable treatment under the Geneva Conventions or the Additional Protocol, then they must be "treated humanely in all circumstances." These provisions establish obligations to take account of others' interests, including the human dignity of enemy combatants. Use of autonomous weapons to kill "wrongdoer" human targets completely bypasses such obligations and represents a modern-day example of Kant's "disgraceful punishments" amounting to "outrages upon personal dignity."

Manipulating a human agent's thoughts so as to distort their freedom of expression, beliefs, and actions would not respect rights to freedom of expression, thought, conscience, and religion. As was alluded to in the EU AI Guidelines, this could impact on participation in political processes and voting rights.

The 2020 UN, OSCE, and OAS Joint Declaration on Freedom of Expression and Elections in the Digital Age notes the alarming misuse of social media by both state and private actors to subvert election processes, including through various forms of inauthentic behaviour and the use of "computational propaganda" (employing automated tools to influence behavior).[13] Recommendation 1(a)(i) requires states to have in place a regulatory and institutional framework that promotes a free, independent and diverse media, in both the legacy and digital media sectors, which is able to provide voters with access to comprehensive, accurate, and reliable information about parties, candidates, and the wider electoral process. Thus, the state has a duty to protect human agents from manipulation by AI systems by providing a legal framework within which such systems will operate, and regulating the conduct of third party actors. This is similar to the German constitutional law state duty to refrain from acts that violate human dignity, and to protect individuals against harmful acts of third parties.

Recommendation 2(a)(ii) requires non-state actors, such as digital media and platforms companies, to make a reasonable effort to adopt measures that make it possible for users to access a diversity of political views and perspectives. In particular, they should make sure that automated tools such as algorithmic ranking, do not, whether intentionally or unintentionally, unduly hinder access to election related content and the availability of a diversity of viewpoints to users. The nature of actions to be taken by private companies is somewhat weakened by the wording "reasonable effort" and "unduly hinder." They should make a "reasonable effort" to adopt measures that make it possible for users to access a diversity of political views and perspectives.

But what constitutes a "reasonable effort"? For this to have any practical meaning and impact on potential users, the digital media/platform company would need, for example, to ensure that: it does not prioritize news outlets from whom it receives advertising revenue or direct funding; small, independent, or foreign news outlets are accessible and not blocked; and users are able to clearly see how to access a variety of sites. "Reasonable effort" could not be discharged by pointing to the comments section as representing a diversity of views and perspectives.

In making sure that automated tools do not "unduly hinder" access to election related content and the availability of a diversity of viewpoints to users, companies may argue that there is no undue hindrance where the information transmitted is blocked on the grounds of preventing crime or legal censorship. But this would need to be tested against the source(s) of information justifying the grounds for blocking, and be balanced against public interest access to information, freedom of expression, and the freedom of the press to undertake investigative journalism.

For example, Google's announcement that it would de-rank Russia Today and Sputnik falls foul of the diversity principle and engages in censorship to manipulate public opinion ("Google to 'de-rank' Russia Today and Sputnik," 2017). Ironically, the stated aim was to prevent misinformation yet by acting as gatekeeper and arbiter of information the corporate entity is engaging in distortion of information which undermines the human agent's autonomy and rational capacity to decide for themselves. Similarly, Facebook's manipulation of news feeds, as part of a psychology study to alter the emotional content of users' posts, treated users as mere means and undermined their agency to engage in free-thinking and expression (Goel, 2014).

Recommendation 2(a)(v) also requires digital actors to be transparent about the use and practical impact of any automated tools they use, including data harvesting, targeted advertising, and the sharing, ranking and/or removal of content, especially election-related content. Thus, at a minimum, companies deploying AI systems must make sure that potential users are made aware of the various uses and impact of automated tools. However, this requirement is caveated by the phrase "albeit not necessarily the specific coding by which those tools operate" so that, unlike the GDPR, there is no automatic right for users to access information on the logic of an automated decision, or information on the significance and envisaged consequences of automated decision-making. This points to the divergence and complexity of non-legally binding and legally-binding rules across different regional instruments alluded to earlier.

4.2.2 Respecting AI Limitations

AI systems should be designed and developed in such a way that recognizes and respects their own limitations (e.g., lack of pure reasoning; undesirability of deployment in certain contexts). This is a means of controlling what AI systems are used for and maintaining human agency. AI systems should not assume they are interacting with inanimate and determinative objects that simply require binary responses. The lack of reflective judgement in machines and algorithms is clearly a handicap for many situations in which answers are less clear-cut, exploratory rather than determinative in nature, and need time for deliberation. A certain level of exchange and interaction is needed to gauge what may or may not be appropriate. A person may be unsure about their options, implications of taking a particular option, their ability to change options later, the consequences from this, or about how the AI system will retain and use their personal data. In accordance with rights under the GDPR and Convention 108+, recourse to a human should be available in such circumstances.

An example of problems relating to respecting AI limitations is an AI-based decision support system designed to certify a person's health status and free movement. The system may have pre-programmed biases that cause rejection of individuals from certain locations or postcodes with high COVID-19 reproduction rates or that are in lockdown. Although the biases are a result of human error, the fact that the system is being deployed in a scenario with a high-risk of detrimental outcomes for personal freedoms points to the need to set and respect limitations.

Even a human reviewing the refused automated certification may be susceptible to automation bias so that there is minimal exercise of reflective judgement and over-reliance on the AI system's decision that the application should be rejected. The combination of AI and human deficiencies means it may be better to rely completely on an accountable human decision-making process rather than a human-machine process from which it may be difficult to untangle errors, attribute responsibility, and seek redress.

4.2.3 Respectful Prioritization of Human Needs

As already mentioned above under agent autonomy recognition, there should be a "human-centric approach" to AI design and development that prioritizes human wants, needs, and values. More specifically, the AI system should enable human agent preferences and choices to curtail its application and use.

Van Kleek et al. (2018) refer to "obstacle respect" in an AI system that understands human agents in a particular way in order to pursue its own goals (e.g., treating the customer as a product, either through advertising or data mining in order to generate profit). Such a system would treat the human agent as a means to an end, and prioritize profit generation over data protection, privacy, autonomy, and rational capacity. Unless the human agent is informed of such prioritization and consents beforehand, it would undermine both the status and respectful treatment dimensions of human dignity.

Another means of respecting prioritization of human needs is to allocate the distribution of harm resulting from the AI system. Prior to purchase and engagement with the system, there should be full and clear disclosure of how the AI system will distribute harm in unavoidable harm scenarios. For example, in the case of autonomous vehicles, will these be designed to prioritize avoiding injury to drivers, pedestrians, or other drivers? Or will they adopt a utilitarian approach to minimize casualties and maximize lives saved? Consumers may be reluctant to purchase autonomous vehicles that fail to prioritize the driver's safety, so some car manufacturers have already declared prioritization of driver safety (Taylor, 2016).

5 Conclusion

In positioning ourselves ready for the "fourth revolution," we should understand the crisis of the self brought on by the AI innovation Ferris-wheel and recognize human dignity represents an important moral and legal norm to focus policy formulation and laws governing AI innovation and impact on societies.

Diverse cultures, international legal instruments, and constitutional laws represent human dignity as a universal moral value and, in the case of German and South African constitutional laws, a right and a duty. Kantian deontological ethics provides the most rigorous exposition of human dignity with its consideration of innate human worthiness and rational capacity. It is from these human characteristics that we can begin to understand human dignity as meaning recognition of the status of human beings as agents with autonomy and rational capacity to exercise reasoning, judgement, and choice; and respectful treatment of human agents so that their capacity is not diminished or lost through interaction with or use of the technology. Human dignity means innate human worthiness that justifies the autonomy and rational capacity status of human agents, and their respectful treatment.

AI that diminishes human agency to exercise reasoning, judgement, and choice undermines human dignity. Part of the problem stems from a sense of losing control over what human activities will be overtaken or replaced by the technology, and its subsequent impact on human autonomy and rational capacity, which relate to the status aspect of human dignity. Several measures can be adopted to protect human autonomy and rational capacity.

First, a "human-centric approach" to regulation whereby AI design and development prioritizes human wants, needs, and values, as well as user awareness and rights. Second, representation of non-deterministic influences on decision-making in AI systems, and if this is not possible to inform users of the AI's limitations in terms of deterministic decision-making and allow them to opt out. Third, acceptance and safeguarding of human pure reasoning; that is, the ability to engage in deep and critical reflective thinking to mitigate challenges posed by AI.

Reflective rather than determinant judgement differentiates humans from machines and algorithms. Without pre-programmed or preexisting rules, humans are able to reconcile and calibrate understanding and reasoning in order to reach a judgement. In the exercise of choice, human agents should be provided with alternatives if they do not want to use AI for resources or services, including prior to use and during use, and for the opportunity to opt out or withdraw consent.

Respectful treatment of human agents so that their capacity is not diminished or lost through interaction with or use of the technology involves three main categories of policy and legal requirements. First, is to respect human agent rights. Designers and developers of AI systems will need to adopt a "human-centric approach" that takes account of rights to privacy, data protection, and fundamental rights.

The GDPR and Convention 108+ provide for extensive privacy and data protection rights throughout the lifecycle of a system, ranging from prior consent for automated decision-making to its contestation. Consideration of fundamental rights to freedom from torture, cruel, inhuman or degrading treatment; freedom of expression; freedom of thought, conscience, and religion; and freedom of association, may necessitate policy decisions not to deploy AI systems in certain circumstances (e.g., warfare; where the AI system can manipulate a human agent's thoughts to distort their freedom of expression, beliefs, and actions, or participation in political processes and voting rights).

The Joint Declaration on Freedom of Expression and Elections in the Digital Age recognizes particular harmful effects from automated tools used to influence public opinion and access to media, and seeks to include both state and non-state actors in a regulatory framework.

Second, recognizing the limitations of AI systems (e.g., deterministic decision-making; lack of pure reasoning) may lead to early policy decisions not to deploy them in open-ended, uncontrolled circumstances where there is a high risk of detrimental outcomes to humans. This can be due to the need for human deliberation, interaction, and reflective judgement (e.g., user's uncertainty about options and their implications; AI-based decision support system certifying a person's health status for free movement purposes).

Finally, respecting prioritization of human needs can be achieved by adopting a "human-centric approach" to regulation and design of AI systems to prioritize information about the system, consent, and allocation of distribution of harm as key features.

Notes

1 "Then God said, "Let Us make man in Our image, according to Our likeness ... So God created man in His own image."
2 "In the day that God created man, He made him in the likeness of God."
3 "Whoever sheds man's blood, By man his blood shall be shed; For in the image of God He made man."
4 *Ibid*, 90–91, paragraphs 64–66 [428–429]; 101–102, paragraphs 87–88 [440]; 96–97, paragraphs 77–79 [435–436].
5 Bundesverfassungsgericht 125, 175, Judgement of 9 February 2010; Bundesverfassungsgericht 132, 134, Judgement of 18 July 2012. See also, Grimm (2013).
6 Bundesverfassungsgericht 87, 209 (228), Order of 20 October 1992; Bundesverfassungsgericht 115, 118 (152), Judgement of 15 February 2006.
7 See also Kant's noumenal / phenomenal world distinction (1781/1998, pp. 338–353).
8 Article 5, Universal Declaration of Humans Rights (UDHR); Article 5, African Charter on Human and Peoples' Rights (AfCHR); Article 5, American Convention on Human Rights (AmCHR); Article 27, American Declaration of the Rights and Duties of Man (AmDR); Article 8, Arab Charter on Human Rights; Article 3, European Convention on Human Rights (ECHR); Articles 4, 7, and 10, International Covenant on Civil and Political Rights (ICCPR).
9 Article 19, UDHR; Article 19, ICCPR; General Comments 10 [19] (Article 19) and 11 [19] (Article 20) of the Human Rights Committee (CCPR/C/21/Rev.1 of 19 May 1989); Article 9, AfCHR; Article 13, AmCHR; Article 10, ECHR.
10 Article 18, UDHR; Article 18, ICCPR; Article 9, ECHR.
11 Article 20(1), UDHR; Articles 21 and 22, ICCPR; General Comment 25 (Article 25) of the Human Rights Committee (participation in public affairs and the right to vote); Article 8, International Covenant on Economic, Social and Cultural Rights; Articles 10 and 11, AfCHR; Articles 21 and 22, AmDR; Articles 15 and 16, AmCHR; Article 11, ECHR.

12 The majority decision held that Common Article 3 expresses "minimum rules applicable to international and non-international conflicts" (paragraph 219), and these rules reflect "elementary considerations of humanity" (paragraph 218).
13 The United Nations Special Rapporteur on Freedom of Opinion and Expression, the Organization for Security and Co-operation in Europe Representative on Freedom of the Media, and the Organization of American States Special Rapporteur on Freedom of Expression, Joint Declaration on Freedom of Expression and Elections in the Digital Age, 30 April 2020.

References

An'Xian, L. (2014). Human dignity in traditional Chinese confucianism. In M. Düwell, J. Braarvig, R. Brownsword, and D. Mieth (eds.) *The Cambridge handbook of human dignity*, pp. 177–181. Cambridge University Press.

Aquinas, T. (2006). *Summa Theologica, Part II-II (Secunda Secundae)*. (Fathers of the English Dominican Province, Trans.) Project Gutenburg. http://www.gutenberg.org/cache/epub/17611/pg17611-images.html (Original work published 13th century).

Braarvig, J. (2014). Hinduism: The universal self in a class system. In M. Düwell, J. Braarvig, R. Brownsword, and D. Mieth (eds.) *The Cambridge handbook of human dignity*, pp. 163–169. Cambridge University Press.

Bundesverfassungsgericht [Federal Constitutional Court]. (2000). 1, 97 (104) Order of 19 December 1951; Bundesverfassungsgericht 102, 347 (367), Judgement of 12 December 2000.

Buolamwini, J., & Gebru, T. (2018). Gender shades: Intersectional accuracy disparities in commercial gender classification. *Proceedings of Machine Learning Research, 81*, 1–15.

Constitutional Court of South Africa. (2000). *Dawood, Shalabi and Thomas v Minister of Home Affairs*. 2000 (3) SA 936 (CC).

Constitutional Court of South Africa. (2001). *South Africa v Grootboom*. 2001 SA 46 (CC).

Constitutional Court of South Africa. (2004). *Khosa v Minister of Social Development*. 2004 (6) SA 505 (CC), paragraph 41.

Council of Europe Modernised Convention for the Protection of Individuals with Regard to Automatic Processing of Personal Data (ETS No. 108+), Amending Protocol to the Convention, adopted by the Committee of Ministers at its 128th Session in Elsinore on 18 May 2018.

Cummings, M. (2004). Automation bias in intelligent time critical decision support systems. *Collection of Technical Papers—AIAA 1st Intelligent Systems Technical Conference*. DOI 10.2514/6.2004-6313.

DeStefano, D., & LeFevre, J. A. (2007). Cognitive load in hypertext reading: A Review. *Computers in Human Behavior, 23*(3),1616–1641.

English Standard Version Bible. (2001). ESV Online. https://esv.literalword.com/.

Errattahia, R., El Hannania, A., & Ouahmanewe, H. (2018). Automatic speech recognition errors detection and correction: A Review. *Procedia Computer Science,128*, 32–37.

European Parliament and Council of the European Union. (2016, April 27). *Regulation 2016/679 on the protection of natural persons with regard to the processing of personal data and on the free movement of such data, repealing Directive 95/46/EC* (General Data Protection Regulation) (GDPR), (OJ L 119, 4.5.2016).

European Parliament and Council of the European Union. (2021, April 21). *Proposal for a regulation laying down harmonised rules on artificial intelligence (Artificial Intelligence Act) and amending certain Union legislative acts*, (COM(2021) 206 final).

Federal Republic of Germany. (1949). German Basic Law, Basic Law for the Federal Republic of Germany in the revised version published in the Federal Law Gazette Part III, classification number 100–1, as last amended by Article 1 of the Act of 28 March 2019. *Federal Law Gazette I*, 404.

Floridi, L. (2014). *The fourth revolution: How the infosphere is reshaping human reality*. Oxford University Press.

Freeman, C., & Louçã, F. (2001). *As time goes by: From the industrial revolutions to the information revolution*. Oxford University Press.

G20. (2019). *G20 human-centred AI principles*. https://www.mofa.go.jp/files/000486596.pdf.

Gandhi, M. (1948). Letter addressed to UNESCO by Mahatma Gandhi, 25 May 1947. In UNESCO (ed.) *Human Rights: Comments and Interpretations*, Paris, 25 July 1948.

Goel,V. (2014, June 29). Facebook tinkers with users' emotions in news feed experiment, stirring outcry. *The New York Times*.https://www.nytimes.com/2014/06/30/technology/facebook-tinkers-with-users-emotions-in-news-feed-experiment-stirring-outcry.html.

Google to 'de-rank' Russia Today and Sputnik. BBC. (2017, November 21). https://www.bbc.co.uk/news/technology-42065644.

Grimm, D. (2013). Dignity in a legal context: Dignity as an absolute right. In C. McCrudden (ed.) *Understanding human dignity*, pp. 381–391. Oxford University Press.

Han, B. C. (2017). *Psychopolitics: Neoliberalism and new technologies of power*. Verso.

Heuss, T. (1993). 4th session of the Committee for Fundamental Constitutional Questions on 23 September 1948. In D. Bundestag and Bundesarchiv (eds.), *Der Parlamentarische Rat*, vol. 5/I, p. 72. Harald Boldt Verlag.

High-Level Expert Group on Artificial Intelligence, European Commission. (2019). *2019 EU AI guidelines—Ethics guidelines for trustworthy AI*.https://ec.europa.eu/futurium/en/ai-alliance-consultation/guidelines/2.

IEEE Ethically Aligned Design for Autonomous and Intelligent Systems—The IEEE Global Initiative on Ethics of Autonomous and Intelligent Systems. (2019, April 4). Ethically aligned design: A vision for prioritizing human well-being with autonomous and intelligent systems. https://ethicsinaction.ieee.org.

Iglesias, T. (2001). Bedrock truths and the dignity of the individual. *Logos: A Journal of Catholic Thought and Culture*, 4, 114–134.

International Court of Justice. (1986). Military and Paramilitary Activities in and against Nicaragua (Nicaragua v. USA), Merits Judgment of 27 June 1986. *ICJ Reports*, 14.

Kant, I. (1969). *The moral law: Kant's groundwork of the metaphysic of morals*. (H. J. Paton trans.) Hutchinson & Co. (Original work published 1785.)

Kant, I. (1996). *The metaphysics of morals*. (M. Gregor trans. and ed.) Cambridge University Press, pp. 105–109, 209–210, 211–213. (Original work published 1797.)

Kant, I. (1998). *Critique of pure reason*. (P. Guyer and A. Wood trans.) Cambridge University Press. (Original work published 1781.)

Kant, I. (2000). *Critique of the power of judgement*. (P. Guyer ed., P. Guyer and E. Matthews trans.) Cambridge University Press. (Original work published 1790.)

Kant, I. (2004). *Prolegomena to any future metaphysics*. (G. Hatfield trans. and ed.) Cambridge University Press. (Original work published 1783.)

Linebaugh, H. (2013, December 29). I worked on the US drone program. The public should know what really goes on. *The Guardian*. https://www.theguardian.com/commentisfree/2013/dec/29/drones-us-military.

Maróth, M. (2014). Human dignity in the Islamic world. In M. Düwell, J. Braarvig, R. Brownsword, and D. Mieth (eds.) *The Cambridge handbook of human dignity*, pp. 155–162. Cambridge University Press.

Metz, T. (2014). Dignity in the ubuntu tradition. In M. Düwell, J. Braarvig, R. Brownsword, and D. Mieth (eds.) *The Cambridge handbook of human dignity*, pp. 310–318. Cambridge University Press.

Norwegian Data Protection Authority. (2020, August 7). *Advance notification of order to rectify unfairly processed and incorrect personal data—International Baccalaureate Organization*.https://www.datatilsynet.no/contentassets/04df776f85f64562945f1d261b4add1b/advance-notification-of-order-to-rectify-unfairly-processed-and-incorrect-personal-data.pdf.

Open Letter by Academics. (2019, July). *Technical flaws of pretrial risk assessment raise grave concerns*. https://dam-prod.media.mit.edu/x/2019/07/16/TechnicalFlawsOfPretrial_ML%20site.pdf?source=post_page.

Organisation for Economic Co-operation and Development. (2019). *OECD recommendation on AI*. https://legalinstruments.oecd.org/en/instruments/OECD-LEGAL-0449.

Power, M. (2013, October 22). Confessions of a drone warrior. *GQ*.https://www.gq.com/story/drone-uav-pilot-assassination.

Pretrial Justice Institute (2020, February 7). *Updated position on pretrial risk assessment tools*.https://university.pretrial.org/HigherLogic/System/DownloadDocumentFile.ashx?DocumentFileKey=417a859-9fe9-2a6c-5e12-3f472d0dc997&forceDialog=0.

Raja, P., & Dietrich, M. (2010). Complacency and bias in human use of automation: An attentional integration. *Human Factors*, 52, 381–410. DOI 10.1177/0018720810376055.

Rosenberg, N., & Birdzell Jr., L. E. (1986). *How the West grew rich: The economic transformation of the industrial world*. Basic Books.

Royakkers, L., & van Est, R. (2010). The cubicle warrior: The Marionette of digitalized warfare. *Ethics and Information Technology, 12,* 289–296.

Semuels, A. (2020, August 6). Millions of Americans have lost jobs in the pandemic—And robots and AI are replacing them faster than ever. *Time Magazine.* https://time.com/5876604/machines-jobs-coronavirus/.

Singer, N., & Metz, C. (2019, December 19). Many facial-recognition systems are biased, says U.S. study. *The New York Times.* https://www.nytimes.com/2019/12/19/technology/facial-recognition-bias.html.

Small, G., & Vorgan, G. (2008). *iBrain: Surviving the technological alteration of the modern mind.* Collins.

Standards Administration of China (2018, January). *China's AI standardisation White Paper—China's White Paper on artificial intelligence standardisation.* 人工智能标准化白皮书（白皮书）（2018年1月，中国标准管理局）.

Stiglitz, J. (2018, September 18). *The future of work.* Lecture delivered at The Royal Society.

Suchman, L. (1985). *Plans and situated actions: The problem of human-machine communication.* Xerox Corporation.

Sweller, J. (1999). *Instructional design in technical areas.* Australian Council for Educational Research.

Taylor, J. (2019, August 14). Major breach found in biometrics system used by banks, UK police and defence firms. *The Guardian.* https://www.theguardian.com/technology/2019/aug/14/major-breach-found-in-biometrics-system-used-by-banks-uk-police-and-defence-firms.

Taylor, M. (2016, October 7). Self-driving Mercedes-Benzes will prioritize occupant safety over pedestrians. *Car and Driver.* http://blog.caranddriver.com/self-driving-mercedes-will-prioritize-occupant-safety-over-pedestrians/.

Tressell, R. (2004). *The ragged trousered philanthropists.* Penguin Randomhouse.

UK House of Lords Select Committee (2018, April 16). *AI in the UK: ready, willing and able?* pp. 317–318. Published by the Authority of the House of Lords (HL Paper 100).

Ulgen, O. (2017). Kantian ethics in the age of artificial intelligence and robotics. *Zoom-in (Questions of International Law), 43,* 59–83.

Ulgen, O. (2019a). Human dignity in an age of autonomous weapons: Are we in danger of losing an 'elementary consideration of humanity'? *Baltic Yearbook of International Law, 17,* 169–196.

Ulgen, O. (2019b). Technological innovations and the changing character of warfare: The significance of the 1949 Geneva Conventions seventy years on. *Journal of International Law of Peace and Armed Conflict* (Humanitäres Völkerrecht), *3–4,* 215–228.

Ulgen, O. (2020). User rights and adaptive A/IS—From passive interaction to real empowerment. In R. A. Sottilare and J. Schwarz (eds.) *HCII Conference Proceedings, in LNCS Series,* 12214, pp. 205–217. Springer.

UN General Assembly (1948, December 10). Resolution 217 A(III), 3d Sess, Supp No 13, UN Doc A/810 (1948).

UN Secretary General's High-Level Panel on Digital Cooperation. (2019). *The age of digital interdependence: Recommendations of the UN secretary*

general's high-level panel on digital cooperation. https://www.un.org/en/digital-cooperation-panel/.
UN Treaty Series. (1966a). *International Covenant on Civil and Political Rights, 999 U.N.T.S. 171* (adopted 16 December 1966, entered into force 23 March 1976).
UN Treaty Series. (1966b). *International Covenant on Economic, Social, and Cultural Rights, 993 U.N.T.S. 3* (adopted 16 December 1966, entered into force 3 January 1976).
Van Kleek, M., Seymour, W., Binns, R., & Shadbolt, N. (2018). Respectful things: Adding social intelligence to smart devices. *Living in the Internet of Things,* 1–6. Department of Computer Science, University of Oxford. https://hip.cat/papers/petras-iet-respectful-things.pdf.
Waldron, J. (2009). Dignity, rank, and rights. *The Tanner Lectures on Human Values,* University of California, Berkeley, 21–23 April 2009.
Weizenbaum, J. (1976). *Computer power and human reason: From judgement to calculation.* WH Freeman & Company.
Woods, C. (2015). *Sudden justice: America's secret drone wars.* Hurst & Co.
Zhu, E. (1999). Hypermedia interface design: The Effects of number of links and granularity of nodes. *Journal of Educational Multimedia and Hypermedia, 8*(3), 331–358.
Zimmermann A. (2020, August 14, updated 2021, September 9). The A-level results injustice shows why algorithms are never neutral. *New Statesman Politics.* https://www.newstatesman.com/politics/2020/08/the-a-level-results-injustice-shows-why-algorithms-are-never-neutral.

Chapter 2

Flipping Rocks and Pointing out the Bugs
Invisible Threats and Data

Darian J. DeFalco

1 Invisible Risks: Retrospective

1.1 Introduction

History is useful to formulate a genealogy of known-risks for any given technology, but data science as it is currently practiced is rather new. Although it has been built out of the elements of earlier discursive formations, it is the combinatorial nature of data science along with the increasing rate of technological performance which creates both its catalytic insights and invisible risks by exposing hidden and unanticipated truths. Even the most concerned individual lacks the ability to track all the potential ethical risks inherent in combining new sets of data to novel analytic processes with massive computational resources available on-demand. Though the risks generally fall under specific areas of concern, such as privacy, fairness, and security, how these problems emerge can be elusive for many reasons.

With the explosion of cloud providers along with the maturity of data pipelines optimized for computational concurrency, the ease of quickly building massive systems to analyze data as needed, the limitations of even Moore's Law no longer apply as the single limiter of data analytic performance. The capacity to acquire, store, analyze, and publish massive sets of information has gone from an expensive organizational commitment in infrastructure and personnel, to a spontaneous opportunity accessible to the public at large. Twenty years ago, the resources required to perform such demanding tasks were largely the domain of government agencies, academia, and well financed corporations requiring massive investments in infrastructure. Now an entire high performance computational cluster can be created in a matter of minutes, on demand, then simply destroyed when no longer needed, paying only for the time in use.

This on-demand model of computing has enabled more participants in the field of data science and AI than ever before possible, ushering in an age of increased use of data science and AI beyond their original tasks (e.g., census taking) to applications both exotic—such as identifying risk in decaying urban infrastructure—and pedestrian—such as making music recommendations in automatically generated playlists. Companies, government agencies, and researchers are more frequently using these tools to address an increasingly broad range of problems.

Mindful of both responsibility and liability, companies and organizations have naturally tried to seek out a clearly delineated list of specific risks or risk domains, with the intention of using them to address the need of legal compliance, ethical concern, and potential liability. Many of these problems are commonly understood and subject to increasing levels of both scrutiny and education, such as privacy and the range of problems that fall under just that one domain of concern. Though these actions were prudent to avoid known problems and risks, a survey of risks and how they were identified reveals that they are frequently the result of well-intentioned actors (as well-informed as could be expected) encountering hidden risks not previously encountered or anticipated.

1.2 IBM

As a field, data science did not announce its own birth. Yet the tools of data science can arguably be linked back to IBM's precursor to the modern computer: Hollerith tabulation machines. The original application of Hollerith cards goes back to the US census of 1890, after the previous US census of 1880 took ten years to complete. Hollerith's efficiency was proven when the 1890 census was completed two years ahead of schedule. This led to other countries purchasing Hollerith equipment for censuses, including the German state of Prussia. IBM had acquired a German company that sold tabulation equipment to the German government.

When Hitler came to power, Germany passed a law that no senior management of a company can be Jewish. With few exceptions (notably Warner Bros.). American businesses operating in Germany accepted these laws and complied with them. Thomas Watson, at the time both CEO of IBM and President of the International Chamber of Commerce, expressed a belief that the role of business and commerce transcended the judgement of laws. Watson was quoted at the time as saying "I'm an internationalist. I cooperate with all forms of government, regardless of whether I can subscribe to all of their principles or not" (Jones & Brown, 2019). Watson notably met privately with Adolf Hitler in his role as President of the ICC, sending a letter afterward thanking Hitler for the high honor, and assuring that he would "endeavor to do all in my power to create more intimate bonds between our two great nations" (Murphy, 2019).

Without speculating about Watson's personal ethical considerations, it is important to note that not only were his actions legal, but they were largely consistent with other American multinational corporations at the time. His role as CEO made ethical considerations for IBM, its employees and partners, as well as the individuals whose lives this technology impacted. Tabulation equipment was not merely used adjacent to the goals of streamlining tracking of Jewish individuals—they were catalytic in enhancing the efficiency of these efforts. Watson was surely aware of the controversy surrounding Hitler and his public position regarding the Jewish population. IBM may not have been aware of how the Hollerith machines were precisely being used by the German government, but they continued to invest in their German operations by building a factory there to speed up delivery time.

Nazi Germany was not the only country with an oppressive human rights record to whom IBM sold tabulation equipment. IBM cooperated with South African Apartheid until 1952 with the sale of an electronic tabulator. This is consistent with the stated policy of Thomas Watson, who felt business and trade were a greater good, irrespective of how the goods or services being traded are applied (IBM, ND).

Given the potential applications of the goods and services in question—notwithstanding their benign intended use—it may not be clear whether selling equipment that can help track, segregate, and oppress is essentially the same ethical consideration as selling goods or services of a more banal nature. That lack of clarity coupled with the potential for extreme outcomes suggests possible ethical disagreement, that is, a situation which may yield different perspectives from the people in a position to make ethical decisions and from those who were implicated or impacted by them. One could argue that Watson was consistent in his bias to dismiss any proximal moral or ethical role, but his willingness to ignore these ethical issues deprived those who worked for him of their own ethical agency. Independent of Watson's personal ethical choices—but directly a consequence of them—individuals facilitated by the efficiency of Hollerith cards transported millions to gas chambers.

A Polish Railways employee who worked in the Hollerith Department was not aware that his office was involved with transporting those deemed undesirable to gas chambers—he just knew it made controlling Polish railway traffic possible. Though Polish Railways did not work for IBM, and Watson was most likely unaware of how the equipment was being used, Watson's failure to consider potential implications of technology under his control started a sequence of events with ethical implications for many involved who were not fully informed (nor in a position to be).

Malicious intent cannot reasonably be attributed to Watson's choices (from the information available), but by enabling the availability of these powerful tools, his choices had an impact on many of these outcomes. Tabulation technology, even in its origins, introduced a level of increasingly efficient enumerative precision, which made previously daunting tasks achievable in record time. That simple act of counting and tracking is on its surface harmless. But a failure to contextualize and evaluate possible uses by malicious actors paved the way for horrendous humanitarian abuses, including the Nazi's system of death camps and South African Apartheid.

1.3 Henrietta Lacks

In 1951, at the age of 31, Henrietta Lacks, a poor tobacco farmer, died from cervical cancer at Johns Hopkins, one of the few hospitals that treated both white and black patients in Baltimore. After she died, scientists extracted, retained, and reproduced cells from her cervical cancer tumor, despite the fact that neither she nor her family ever gave permission (Callaway, 2013). These "HeLa" cells represented the first time that scientists were able to continuously grow human cells in a laboratory. This represented a major breakthrough in research, allowing scientists to share these cells widely, helping the development of the polio vaccine, and informing over 60,000 research papers.

Regardless of all the benefits derived from this source of genetic information, it was not until August of 2013 when NIH reached an agreement with the descendants of Henrietta Lacks, defining a clear process on how Henrietta and her ancestor's genetic property may be used going forward. Her family was not even aware of the existence of the HeLa cell line until 1973 when scientists contacted them in order to collect blood samples (Zimmer, 2013). Several months prior to NIH's agreement with Henrietta Lacks' ancestors, researchers from the European Molecular Biology Laboratory (EMBL) published a full sequence from a specific strain of the HeLa cell line on the Internet in 2013—without consent from her family, as was the common practice at the time. Once her family became aware, the scientists complied with the family's wishes that her genomic data be taken offline. However, this was done well after the data had been readily available and widely downloaded (Greely & Cho, 2013). Technically, the scientists at EMBL followed all applicable laws and regulations, as the privacy issues involved were not sufficiently well understood nor treated with sufficient caution. Their intention behind sharing the full sequence was to provide a resource for other researchers. This priority clearly outweighed any concern about widely publishing personally identifying information, absent any form of safeguarding or even notification, let alone informed consent.

The invisible risks behind sharing this data publicly, and how it compromised the rights of Henrietta's family, subsequently exposing an atlas of genetic faults and liabilities to the world by freely publishing them, were not well understood at the time. Publishing her DNA sequence was indeed a significant escalation of this risk, but the original violation of sharing her cells predated the ability to carry out such a dangerous action by over half a century. If the breadth of risk inherent in publishing the DNA sequence was only then revealing itself, then consider how truly unimaginable it would have seemed to those researchers in Baltimore in 1952.

2 Invisible Risks: Contemporary

2.1 Identifying Individuals Using DNA

The fallout from the publication of Henrietta Lacks DNA sequence was infamous insofar as it ushered in procedures and measures to prevent similar events from reoccurring. Common technical solutions both then and now involved expensive investments in various controls, used to restrict access and track usage. Aside from their cost (which often consumes sizable chunks of precious research budgets) these systems also create additional procedures, slowing down research, layering on needs for additional personnel to support complex technical infrastructure.

Alternative strategies used to secure genomic data include obscuring sources by ensuring data has been stripped of any personally identifiable information, commonly referred to as "security through obscurity." On the surface, it appeared that this obscured set of genomic data (whose audience at the time was limited largely to researchers) was reasonably safeguarded (Narayanan & Shmatikov, 2019). Yet, in 2014, Erlich and Narayanan published research that argued if a publicly searchable site were to grow large enough, searching for the Y chromosome, 12% of males would be traceable by inferring their surname. In 2018, exactly as Erlich and Narayanan predicted, investigators were able to track the Golden State Killer using a sample of his DNA and the public website *GEDmatch*. Using *GEDmatch*, police were able to take a sample from the crime scene, sequence the DNA, upload it as if they were somebody looking for relatives, and trace the DNA back to the Golden State Killer by matching it to one of his relatives.

Even though *GEDmatch* is a smaller service than larger services such as *23andme* or *ancestry.com,* because *GEDmatch* permits third party DNA to be uploaded by groups trying to find a match, it enables anybody with access to a male's DNA, to sequence it, upload it, and use it to identify and find a person. That same year, Erlich and Narayanan determined that *GEDmatch* at the time had at least a third cousin match for 60% of European Americans, and within three years, would be sufficiently large to find virtually any European American this way. Whereas in the case of Henrietta Lacks' ancestors, privacy was compromised because researchers exposed their DNA without consent, this later exploit (by a government-sanctioned entity) is not the consequence of third parties misusing data to which they did not have legal access. Instead, it was enabled by the choice of individuals to share their DNA to further genealogical pursuit, a seemingly benign and commonly practiced activity.

Irrespective of the impact on personal freedoms from individuals being potentially pursued and apprehended using this method—or individuals sharing their genomic information freely—the implication for genetic researchers who regularly deal with sequenced data is even more significant, as it raises the stakes for securing privacy. Security through obscurity is no longer a viable method of securing human genomic data if it is known that the subjects in question can have their identities reverse engineered with commonly available public resources (Narayanan & Felten, 2014).

2.2 Private Traits Using Digital Records

In addition to being able to track people by their DNA, research also has shown how people can be tracked by private actions they take online. Narayanan and Shmatikov (2008) showed how they could identify an individual Netflix subscriber's history in an anonymous dataset of 500,000 subscribers with a minimal amount of information. From this information, they could infer seemingly unrelated information about the individual, such as their political leanings. Kosinski, Stillwell, and Graepel (2013) proved that using just an individual's public Facebook "likes" history, they could detect personal information, including characteristics historically linked with discrimination, with a high level of confidence. Their model correctly discriminated between homosexual and heterosexual men 88% of the time, black from white Americans 95%, and between Democrat and Republican 85%.

The very nature of Facebook likes are public expressions of approval, often to seemingly innocuous concerns. However, there is a good likelihood that people expressing their preferences through Facebook likes are not protecting these expressions in the same way they may with explicit personal details. Kosinski, Stillwell, and Graepel (2013) were trying to build a personalized recommendation system for individuals using a model they developed. The intention was to provide more useful recommendations by reflecting the individuals' detected interests and preferences. The researchers drew a sharp distinction between recorded and statistically predicted behavior, recognizing that the benefits of personalized recommendations may have sharply negative effects, such as predicting that an unmarried woman is pregnant in a culture where this is not acceptable (Kosinski et al., 2013).

2.3 AI and Racial Bias

With the emergence of commercial applications for facial recognition, AI is being used in public and private settings to track, surveil, and monitor people—but not equally. Buolamwini (2017) conducted a comparative study of automated facial recognition machine learning algorithms and found that they encountered error rates up to 34.7% when attempting to detect dark-skinned females. She attributed this to the fact that the composition of the datasets used for two facial analysis benchmarks were between 79.6% and 86.2% of lighter-skinned individuals (Buolamwini, 2017). Buolamwini stumbled upon this implicit bias when she observed that these systems were not readily detecting her face.

It is unreasonable to expect end users of such complicated, cutting-edge tools to understand how algorithms to detect faces are limited by sampling data in the formative stage. A considerable knowledge gulf exists between these users and data scientists or AI engineers who may more readily recognize the types of errors to which systems may be prone. However, even that recognition depends on awareness of omissions in the datasets used to define the models. If non-technical, operational personnel observed high precision when applied to faces that happen to be represented by their training models, an initial encounter with a false identification may not stand out as such. The idea that a machine might have a racial bias that we typically attribute as distinctly human may seem like a foreign concept. This mismatch of expectations and functionality increases the possibility of systemic false identification going unnoticed, creating a powerful risk.

What makes Buolamwini's insight so concerning is that the limitations she identified were not in models for specific algorithms or implementations, but in benchmarks which are used to tune, optimize, and validate the algorithms across multiple products and applications. Because of the sheer size of these datasets, it is highly unlikely that the average user of these benchmarks would perceive an obvious or intrinsic problem by looking at them. The training dataset, created by Intelligence Advanced Research Projects Activity and National Institute of Standards and Technology, consists of "138,000 facial images, 11,000 videos, and 10,000 non-face images" (NIST, 2019). Given the integrity of the source, the size, variety, and seeming completeness of the dataset, the ability to detect or quantify any disproportionality within the dataset would be highly unlikely. That detection would require substantial expertise and manual labor, or else an optimized identification algorithm already known to have exceedingly high reliability (for comparison).

As a result of her research, Buolamwini has persuaded companies to pause facial recognition technology, in order to better understand these biases. Were it not for her own personal intersection with these biases, it is unclear if they would have been so readily detected.

3 Solution Space: Going Forward

3.1 Institutional Power Structures

Understanding new technologies at a sufficient depth to identify and avoid ethical risks needs a clearly defined and articulated process, with accountability attached. Formal processes commonly exist to protect those who raise alarms from potential institutional resistance or apathy that shape the invisible rules of discourse surrounding ethical responsibility. In his *Discourse on Language*, Foucault (1972) describes the way discourse forms invisible systems of exclusion and control which emerge, restricting what is permissible speech. Notably, he identifies two systems of exclusion: permissions and "folly." Permissions would be formal rules, whereas folly would be implicit social pressure. AI systems are visible in settings both private and public, academic and commercial. Whether these systems take the form of explicit rules or informal conventions of behavior, such as professional isolation or obstructing personal goals, as a society we have evidence that institutional power dynamics are well understood.

The invisible risks present within the work of data science and AI usually occur on a large scale. When these have been coupled to commercial interests (e.g., IBM's problematic business partnerships), precedent demonstrates that free-trade and fiduciary responsibilities to shareholders may be identified as conflicts with potential ethical consequences. In matters of liability, companies will generally act prudently to mitigate clearly understood, well-defined risk. However, in cases where the risk may not be obvious—or if its resolution represents a business impact—the danger that it does not receive proper attention is significant. The individual data scientist or engineer may perceive a risk, but the more technical or nuanced the nature of the problem (i.e., obscure to those in positions of authority), the more likely concerns raised might result in either pushback, social judgement, or outright dismissal. This institutional resistance is not specific to data science or AI, but it is a vital dynamic to understand in the context of the types of problems which may arise and the need for clear procedures to ensure these concerns can be addressed safely.

3.2 Institutional Power Structures

Education and training are vital parts of making progress along the lines of promoting consistent and accurate understandings, enabling coherent discussion for all the relevant stakeholders. However, education and training only serve to support and enhance processes that need to be formally defined and embraced across organizations to ensure the imagination required to perceive and identify these risks is not wasted by lacking the proper agency and outlet. Dewey (1920) grappled with many of these issues in framing democracy as a broad mode of activity. While democracy may feel like a poor fit for all situations, democracy for Dewey

> developed only when its elements take part in directing things which are common, things for the sake of which men and women form groups—families, industrial companies, governments, churches, scientific associations and so on. The principle holds as much of one form of association, say in industry and commerce, as it does in government.
> (pp. 199–200)

Rather than a vague singular abstraction, Dewey extends the idea of society as covering "all the ways in which by associating together men share their experiences, and build up common interests and aims; street gangs, schools for burglary, clans, social cliques, trades unions, joint stock corporations, villages and international alliances" (p. 200). Dewey argued the following regarding the open free exchange of ideas:

> The best guarantee of collective efficiency and power is liberation and use of the diversity of individual capacities in initiative, planning, foresight, vigor and endurance. Personality must be educated, and personality cannot be educated by confining its operations to technical and specialized things, or to the less important relationships of life. Full education comes only when there is a responsible share on the part of each person, in proportion to capacity, in shaping the aims and policies of the social groups to which he belongs.
> (p. 209)

It is that inclusive dialectic that has the potential to cut across institutional silos, allowing the professional imagination of an organization's most talented individuals to share the risks they perceive without fear of retribution or dismissal. By providing a formal process where concerns can be shared with relevant stakeholders, the organization has a chance to learn together, and evolve what Dewey might refer to as "their collective mind." Dewey felt the morality of a given act was something specific to the act, so "the good of the situation has to be discovered, projected and attained on the basis of the exact defect and trouble to be rectified." He felt that morality "is not a catalogue of acts nor a set of rules to be applied like drugstore prescriptions or cookbook recipes"; rather, the need is for methods by which "difficulties and evils" can be identified, enabling "plans to be used as working hypotheses in dealing with them" (p. 169).

As organizations grapple with the need to provide clearly delineated lists of known risks, we see examples from notable organizations with ample experience in this field, and how they recognize the need to develop such clear inclusive processes. One such organization is the US Federal Government. In June of 2020, the Office of the Director of National Intelligence released an Artificial Intelligence Ethics Framework for the Intelligence Community. Though this may appear to be an isolated set of considerations, the basic problems and implications that it grapples with have broad relevance to many applications of AI and data science.

Though a high-level framework, it outlines both the wide scope of considerations as well as who should be involved in assessing risks:

> Identifying and addressing risk is best achieved by involving appropriate stakeholders. As such, consumers, technologists, developers, mission personnel, risk management professionals, civil liberties and privacy officers, and legal counsel should utilize this framework collaboratively, each leveraging their respective experiences, perspectives, and professional skills.
>
> (intel.gov, 2020)

4 Conclusion: The Future Is Unwritten

We must aspire to avoid discovering risks by experiencing them. As with Emil Oskar Nobel's ill-fated experiments with nitroglycerin, data science and AI are explosive fields. They are constantly and expanding, driven and perpetuated by insights we gain from them, creating a sustained environment of both nascent and exigent dangers.

This is not a trend which will reverse. Data is what drives discovery. It is the lifeblood of business. As a culture, we have all slowly adapted to the increasing realization that the power of data and the platforms that provide it bring with them a burden of responsibility beyond our initial understanding. Security through obscurity, though popular for many organizations, revealed itself as not a permanent solution, but simply a stopgap measure unsuited for emerging capabilities and data availability. Now with the emergence of new threats such as "deep fake" video manipulation, we should recognize a familiar sense that data once thought harmless—such as a photograph or family movie—could be exploited and weaponized by coupling with readily available and increasingly powerful computational resources.

The element of greatest concern when looking at the risks of data science and AI is the presence of risks that are invisible or seemingly harmless, when in fact they are potent in ways difficult to imagine. The ability to implement data collection, analysis, and manipulation on this scale with this level of performance constitutes an incredibly potent tool which can unintentionally transform abstract pieces of information into tangible, human harm.

Education and deep focus on the underlying issues are necessary but insufficient on their own to safeguard against new and unanticipated risks. The power of these tools coupled with typical institutional power dynamics presents a grave concern that needs to be met by formal, adequately empowered processes, embraced by their parent institutions and involving a cross-section of the relevant stakeholders. We simply must be prepared, with processes in place, to handle whatever bugs we inevitably find when somebody flips another rock.

References

Buolamwini, J. (2017). *Gender shades: Intersectional phenotypic and demographic evaluation of face datasets and gender classifiers*. MIT Master's Thesis. https://www.media.mit.edu/publications/full-gender-shades-thesis-17/.

Callaway, E. (2013). Deal done over HeLa cell line. *Nature, 500*, 132–133.

Dewey, J. (1920). *Reconstruction in philosophy*. Project Gutenberg.

Erlich, Y., & Narayanan, A. (2014). Routes for breaching and protecting genetic privacy. *Nature Reviews Genetics, 15*, 409–421.

Foucault, M. (1972) *The archaeology of knowledge: And the discourse on language*. Pantheon Books.

Greely, H., & Cho, M. (2013). The Henrietta lacks legacy grows. *EMBO Reports*. https://doi.org/10.1038/embor.2013.148.

IBM. (ND). *A history of progress*. https://www.ibm.com/ibm/history/interactive/ibm_history.pdf.

Jones, G., Brown A. (2019) *Thomas J. Watson, IBM and Nazi Germany*. Harvard Business School.

Kosinski, M., Stillwell, D., & Graepel, T. (2013). Private traits and attributes are predictable from digital records of human behavior. *Proceedings of the National Academy of Sciences of the United States of America*, 1218772110.

Narayanan, A., & Felten, E. (2014). *No silver bullet: De-identification still doesn't work*. https://www.cs.princeton.edu/~arvindn/publications/no-silver-bullet-de-identification.pdf.

Narayanan, A., & Shmatikov, V. (2008). *Robust de-anonymization of large sparse datasets*. https://www.cs.utexas.edu/~shmat/shmat_oak08netflix.pdf.

Narayanan, A., & Shmatikov, V. (2019). *Robust de-anonymization of large sparse datasets: A decade later*. https://www.cs.princeton.edu/~arvindn/publications/de-anonymization-retrospective.pdf.

National Institute of Standards and Technology. (2019). Face challenges. *Journal of Research of the National Institute of Standards and Technology*. https://www.nist.gov/programs-projects/face-challenges.

Murphy, H. (2019). Dealing with the devil: The triumph and tragedy of IBM's business with the Third Reich. *The History Teacher, 53*, 171–193. https://www.jstor.org/stable/27058571

Office of the Director of National Intelligence. (2020). *Artificial intelligence ethics framework for the intelligence community*. https://www.intelligence.gov/artificial-intelligence-ethics-framework-for-the-intelligence-community.

Zimmer, C. (2013, August 8). A family consents to a medical gift, 62 years later. *The New York Times*. https://www.nytimes.com/2013/08/08/science/after-decades-of-research-henrietta-lacks-family-is-asked-for-consent.html.

Chapter 3

Seeing the Forest and the Trees

AI Bias, Political Relativity, and the Language of International Relations

Dr. Leah C. Windsor

1 Introduction

Language encodes meaning, and current approaches to event data generation fail to account for the different semantic interpretations of events and event typologies across languages and cultures, and over time.[1] The actions of international agents reflect their preferences (Druckman & Lupia, 2000), which are themselves as subjective as comparative interpretations of color (Churchland & Sejnowski, 1994; Cibelli et al., 2016). Linguistic relativity suggests that the structure of language influences actors' worldview (Whorf, 1940), including grammar and vocabulary that can be modeled computationally using machine learning with multilingual semantic spaces (van Atteveldt et al., 2008; Finch et al., 2005) and word vectors (Mikolov et al., 2013; Pennington et al., 2014).

Addressing AI bias requires an interdisciplinary, integrative approach that leverages the strengths of computer science and computational social science. Biases in machine learning (ML) and artificial intelligence (AI) are well-known, namely that AI systems learn bias from word embeddings (Bolukbasi et al., 2016) and replicate human-like biases (Caliskan et al., 2017) such as gender and racial/ethnic stereotypes (NPR, 2016). This points to a broader and more fundamental source of bias: *the unaccounted changes in meaning across languages, and over time embedded in AI algorithms.* Words and word meanings are not stable across time. They evolve and mutate, adopting new contexts, shedding old interpretations, and sometimes retaining the lingering vestiges of their original intent. This poses conceptual and methodological challenges for studying dynamics of international relations, as the descriptions of political events and those who engage in them are subject to linguistic drifts and shifts.

Scholars study who does what to whom in the international system to uncover trends and patterns, both between countries and over time. These activities are encoded in the language that describes the actions

DOI: 10.4324/9781003030928-5

and the actors. In its most simple form, Source A does Action X to Target A. For example, the leader of a country (Source A) could commit to sending peacekeepers (Action X) to a conflict, or humanitarian aid to another country (Target A). A rebel group could engage in material conflict with another rebel group, or with the central government. A leader could be ousted in a coup by a rival politician. A terrorist group could launch an attack against civilians. Two countries could go to war with each other, or sign a peace treaty to end a conflict. All these activities, and more, are encoded in a methodical system that captures frequently occurring event types. Data generated through this process are called event data, and rely on news reports from major media news outlets like the New York Times, Associated Press, Reuters, and the BBC's Summary of World Broadcasts. The process of event data generation relies on natural language processing programs such as Petrarch and Accent to identify the actors, actions (Schrodt, 2012), and activities, which Schrodt (2017) describes in detail. Fortunately, scholars have bridged many of the ideological and technological gaps previously identified in computational politics (Mallery, 1994).

The two sources of bias I address here rest on two assumptions: first, people from different cultures describe events differently; and second, meaning evolves over time. The business and culture handbook *Kiss, Bow, or Shake Hands* describes in detail how features of negotiation, such as punctuality, gift-giving, partaking in alcohol, and attire, vary across cultures (Morrison & Conaway, 2006). What is appropriate or normal in one country or culture may not translate to other contexts. The same is true with language and events. For example, cattle raids in sub-Saharan Africa are a common indicator of conflict (Butler & Gates, 2012; Eck, 2012), as happened in Sudan and Kenya in 2008 between the Toposa and Turkana groups that killed more than 5,000 cows. In North Korea, on the other hand, military posturing, exchanges of fire, and long-range weapons tests are common activities (Cha, 2002). In essence, what "counts" as conflict varies by culture. Bias resides in the failure to model these cultural and linguistic differences.

2 Conflict

What counts as conflict varies between countries, and it varies over time. Figure 3.1 shows the results of Google N-gram searches for the terms "conflict" and "war" between 1900 and 2018 using the original languages as search terms. The terms war and conflict are semantically similar, and often occur contemporaneously in context. There are several important trends to observe here: in the time span covered, countries follow generally similar time trends for each individual search term, and the trends between war and conflict are distinguishably different.

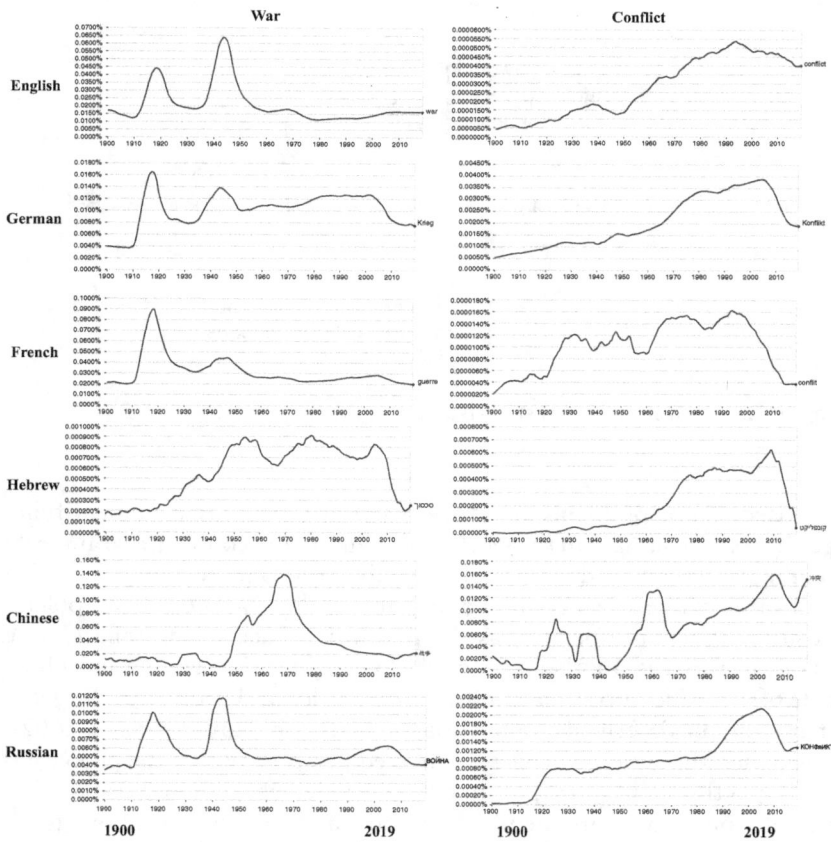

Figure 3.1 Trends in Google N-Gram for "War" and "Conflict" by Language (1900–2018).

In general, the trends for war across language search terms spike around times corresponding to World War I and World War II, whereas the trend for conflict gradually increases over time, rising rapidly from the 1970s onward. For war, however, the Chinese search terms return a substantially different pattern, with little register for the two World Wars, and a large spike around the 1950s corresponding to the Korean War. Additionally, the kurtosis of the graphs differs, with attenuation in the peak for World War II with the French search terms, for example. Conflict, too, shows different patterns over time. In French, the trend reflects a gradual increase over time, while in English and German, the graph jumps suddenly in the 1970s.

If we treat war and conflict as identical concepts across language and over time, we fail to capture the nuance and variation for how these terms

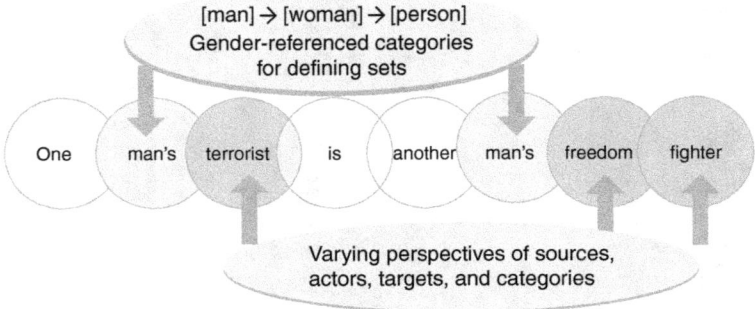

Figure 3.2 Variations in perspectives on political events.

are used. These problems are exacerbated by the fact that in identifying patterns in event data in the world, most automatic coding programs rely on English-language documents. This fails to account for the perspective and variation between countries. Figure 3.2 elaborates on this problem. Position affects perspective, as the political science saying reflects: one man's terrorist is another man's freedom fighter. Depending on which side of a conflict a person stands, *Source A* doing *Action X* to *Target A* may either be described as a terrorist or a freedom fighter (Ganor, 2002).

Further, this axiom encodes a gendered perspective that the people engaging in this activity are presumed to be male, when an increasing number of women who are members of non-state actor organizations (such as terrorist groups) are engaging in extrajudicial, lethal acts of violence. The master codebook for generating events, called CAMEO (Conflict and Mediation Event Observations) contains more than 270 distinct codes for political events in 20 categories that capture a set of the things that happen between countries and groups in the world, which have only recently been translated into Spanish (Osorio et al., 2019). The Western bias is clear in the ordered-ness of the categorical CAMEO codes which exclude some categories (Bagozzi et al., 2019) and which constrain events to discrete values when in reality, the scale is likely continuous (Si & Reiter, 2013).

3 Word Embeddings: Conveyors of Culture

Current approaches for event data generation harness the power of automated content analysis for coding news media reports about world events. Forecasting likely scenarios in volatile and opaque environments itself is precarious; ethnographic approaches, including agent-based modeling, are time-consuming, and big data approaches can aggregate over the essential nuances that often provide meaning and context for consequential events (Ward et al., 2013). A key assumption is that language reflects cultural biases and that, more specifically, such cultural biases are reflected in reports describing major socio-economic-political events of interest to more than one country, nationality, or group. Unsupervised word meaning representations, such as Latent Semantic Analysis and word2vec, that are derived from large collections of documents collected from a specific culture, e.g., Western/American culture, are expected to capture the underlying cultural biases.

Word embeddings are representations of word meanings using vectors of numbers (Antoniak & Mimno, 2018). By representing words numerically, it is possible to calculate the similarity (or dissimilarity) of meaning of words and phrases across documents. Machine learning and deep neural networks generate word embeddings from real information, including the bias that is already encoded in the documents used to train the model (Caliskan et al., 2017). AI algorithms learn gender, racial, and cultural biases because the systems that generate the data used to derive word meanings—i.e., humans—are inherently biased. These biases have real-world consequences in terms of prison sentencing guidelines, resume evaluation for employment, and gender-based discrimination (Greenwald & Krieger, 2006; Leavy, 2018; Zou & Schiebinger, 2018).

Event encoders that are used to classify interactions among global actors along a conflict-cooperation continuum are also used to assign discrete values for sources, actions, and targets—i.e., who did what to whom (Althaus et al., 2017; Lautenschlager et al., 2015; Leetaru, 2014). Yet event data generation does not account for temporal changes in language or geopolitical realities, like a decline in interstate war and increase in civil wars, as well as forms of semantic and cultural bias. Further, static dictionaries discard events if actors are not included, introducing selection bias into the data generating process. As a result, geopolitical event codes reflect a Western/American bias that obscures other cultures' interpretations of events and messages. At present, as previously mentioned, syntactic parsers for event data identify that *Source A* does *Action X* to *Target B*, where X has a static interpretation insensitive to language or time. However, in reality X is highly variable, context-dependent, and language variant. In other words, A interprets X_1 as Y_1Z_1, whereas B interprets X_1 as Y_2Z_2. Bias resides in the semantic difference between Y_1Z_1 and Y_2Z_2.

Changes in word embeddings reflect changes in language usage that suggest differences in culture. A more robust, and linguistically and culturally sensitive system that captures how the embedded word vectors differ between languages and cultures, compared to the standard Western/English-language corpus, will help to reveal culturally-embedded, empirically verifiable language bias. The same event may be reported differently, and though news reports may use similar language to describe events, in fact there are cultural nuances to consider that can be modeled by using culture-specific semantic spaces.

News sources may describe an event using very different language. In the dispute between Russia and Ukraine over contested territory in 2014, Russian news sources initially denied that a military campaign was taking place, and then subsequently said that they were "forced to act." As shown in headlines from varying news sources and countries in Table 3.1, the perspective from Ukraine, the UK, and the US was quite different, describing the event as terrorism and a "land grab." Other research has established that Russian state media, and independent media within Russia, describe political events differently (Labzina & Nieman, 2017), supporting the idea that where you stand influences what you see, and how you talk about it. Though news reports may use similar language to describe events, in fact there are considerable cultural nuances that can be modeled by using culture-specific semantic spaces.

There are currently no multicultural approaches, beyond a translated CAMEO scale for Spanish, for culturally contextualizing event data, which reflect the Western values, priorities, and perspectives that generated the coded data (Bagozzi et al., 2019; Osorio et al., 2019; Solaimani et al., 2016; 2017). The Western bias is clear in the ordered-ness of the categorical CAMEO codes which exclude some categories (Bagozzi et al., 2019) and which constrain events to discrete values when in reality, the scale is likely continuous (Si & Reiter, 2013). This project can potentially add substantial knowledge about bias in event data generation and cross-cultural communication beyond the existing event data bias research.

More broadly, the lack of cultural awareness for machine learning and artificial intelligence is an issue for any AI system whose inputs are

Table 3.1 Variation in news headlines in Russian-Ukrainian conflict

News/Country	Headline
Xinhua	"Putin calls Ukraine events coup, defends Russian position."
New York Times	Russia sent tanks to Separatists in Ukraine, U.S. Says
RT	Putin: Russian citizens, troops threatened in Ukraine, need armed forces' protection
CNN	Report: Ukraine military dolphins to switch nationalities, join Russian navy
New York Times	Obama steps o\up Russia sanctions in Ukraine crisis
Al Jazeera	Ukraine: Lies, propaganda and the West's agenda
Ukraine	Is the Western narrative obscuring what's really going on in Ukraine?
Der Spiegel	Europe toughens stance against Putin
NPR	Putin says those aren't Russian Forces in Crimea
BBC	Nato summit: Alliance 'stands with Ukraine'

from multilingual sources, or even monolingual multicultural sources. To understand the source of linguistic bias, we can align word embeddings (Antoniak & Mimno, 2018; Mikolov et al., 2013) from different languages and sources, and examine the movement of words across different contexts, such as event typologies. Movement of words means changes of context, which means different interpretations of the word. Event data seek to measure who does what to whom, with what effect (Lasswell, 1948), and bias resides in subjective interpretations of events (Bagozzi et al., 2019). Recent NSF-funded research addresses aspects of biases, including the development of tools to integrate documents with different formats with accounts of the same or closely related events (Wu et al., 2019), analytical approaches to capturing correlates of observed events (Nieman et al., 2018a; 2018b), and more representative event data sets that incorporate global and historical sociopolitical events from multiple language sources (Bagozzi et al., 2019). These differences in embedded word vectors represent differences in culture (Gumperz, 1968; Slobin, 1996). In other words, the structure of a language foregrounds various concepts. This is in essence, where culture is found; the nuances of linguistic categories reflect differences in cultural perception.

4 Defining the Levels of Analysis

Bias resides at the semantic, temporal, corporal, source, and cultural levels. Operationalizing culture is daunting, given the more than 6,000 languages in the world, and the many cultures that they represent. One option is to use the World Values Survey Cultural Map which sorts countries into the following cultural categories: English-Speaking, Orthodox, African, Islamic, Latin American, South Asian, Confucian, Protestant Europe, and Catholic Europe. The official languages of the United Nations (English, French, Spanish, Mandarin Chinese, Arabic, and Russian) would serve as an ideal starting point for defining the scope of inquiry for de-biasing news, given that these languages cover a substantial portion of speakers in the world. Culture resides in polysemy, that cultures have different interpretations of the same word, and the same word has many meanings, evidence for which is found in gestures, false cognates, and social expectations (Morrison & Conaway, 2006). Further, there are many ways to describe the same type of event, depending on cultural vantage point and the lexical-semantic perspective a culture holds.

4.1 Case Selection Bias

Researchers may introduce bias into data sets through their choice of cases and sources or through the ontologies that they use to classify the data (Geddes, 1990). However, data sets may also inherit bias from the wider culture. Machine learning algorithms trained on biased data exhibit these same biases; examples have been found in recommender systems for hiring and for parole decisions. These biases can be identified in the purely semantic setting and recent scholarship has tried to identify, quantify, and eliminate these biases in word embeddings (Bolukbasi et al., 2016).

The same tendency of machine learning to reflect wider cultural biases can be used to more effectively identify cultural models. Big data approaches have largely supplanted more time-consuming ethnographic models of social science research for defense-related applications, representing a preference for speedy and agile big data over the slower qualitative, thick data. For example, traditional approaches to agent-based modeling require in-depth interviews to generate effective predictive models (Bueno de Mesquita, 2010). Recent developments in automated coding of news sources, such as ICEWS and GDELT (Lautenschlager et al., 2015; Leetaru, 2014), have simplified event data generation by automating the process (Norris, 2016). By examining the lexical and semantic differences in news media reports about international political phenomena, we can gain a deeper understanding of human beliefs, perspectives, and identities. Without accounting for these biases, our ability to explain and predict social phenomena, and to understand and be understood, will remain unnecessarily hampered.

4.2 Actor Bias

Static dictionary-based named entity recognition programs discard events for which they do not have a source or target actor, which also introduces bias (Solaimani et al., 2016; 2017). A current limitation of Petrarch-2 and other event data generating programs is that they rely on shallow natural language processing and formulaic news formats. This does two things: first, it encourages assigning codes to non-events; and second, it struggles to process non-newswire media reports (Schrodt, 2015). Thus, while the BBC Summary of World Broadcasts and the Foreign Broadcast Information System source articles from several hundred news outlets, current automated event coding filters "local" media.

4.3 Media Bias

Local media reports are ignored because they contain many stories that describe relevant events. Current event coders cannot reliably distinguish these and hence produce false-positive events. This process reflects a Western perspective of what events are important, and how the events are ordered from most cooperative to most conflictual. English is an

SVO (subject verb object) language, and this cognitive construct is encoded in event data programs. Machine learning can help to reveal cultural models of those who designed the ontologies that are used to code those datasets, as there are strong geopolitical and temporal components embedded in these processes.

4.4 Process Bias

Event data encodes cultural values in several ways (Weidmann, 2016). First, the set of events captured by event scores are "tuned" to indicate the categories of interest for the Western scholars that developed those approaches during a particular political era. Interstate political dynamics dominated the Cold War; in the past 30 years conflicts have diversified and become more complex. Recent scholarship has highlighted the promises and problems associated with the rise in automated event data coding (Wang et al., 2016). Event data generation is an ontological question where conflict and cooperation are relationally embedded along a continuum (Saraf & Ramakrishnan, 2016a; Saraf & Ramakrishnan, 2016b). Developers of the coding scheme organized the event codes from least to most conflictual; however, these are subjective decisions that other actors might dispute.

4.5 Temporal Bias

Event data presently encodes temporal bias because it uses contemporary language to assign values to events from the past; language from the 1940s is different from language in 2021, and event types are different as well. During the Cold War, more emphasis was placed on state-level conflicts. In the past 30 years, civil wars and internal conflicts involving multiple non-state actors have dominated the event data space. Related to the missing actor problem, during the Cold War state-based activities held the roster actors relatively stable; in the post-Cold War era, non-state actors have arisen and disappeared with greater frequency. Further, states and non-state actors engage in different types of political activities, notably violence against civilians and sexual and gender-based violence are missing from the coding schemes.

4.6 Semantic Bias

Culture and language are fairly stable but do undergo shifts over time as words and politics evolve. This workflow improves upon existing event data methods by explicitly modeling language over time, and in different cultures. For example, using contemporary language to encode historical instances of conflict and cooperation ignores how word meanings have evolved. The term "conflict" in 1947 versus 2018 would entail a different array of actors, actions, and geographies. Current event coding tools capture only one perspective, as the "meanings" embedded in news media are assigned codes based on the Western/American interpretation of the event. Culture is encoded in language; linguists debate to what extent language shapes or constrains thought and perception (Lucy, 2015; Staff, 2016; Tseng et al., 2016). Shared values, norms, rituals, history, and geography create self-reinforcing group identities that are conveyed through language.

The process of creating a word embedding entails transforming a series of characters into a vector based on its linguistic context. This usage of linguistic context allows semantic relationships to be encoded as co-occurring words often have a semantic relationship; however, a side effect of using linguistic context is the capture of cultural associations and biases that are present in documents themselves. For example, if the corpus used to train the word embeddings often uses words like man in the same context as computer programmer and woman in the same context as homemaker, the program learns gender and other forms of bias. This should produce two distinct but related sources of cultural bias: first, in the corpora used to generate event data; and second, in the codes generated from the corpora. It is clear from the event codes that the codes themselves are biased; for example, sexual/gender-based violence is not coded, because the CAMEO codes were generated by male scholars before gender and sexual violence were mainstream concepts.

We can hypothesize about the other categories of bias we might find in the corpora and event codes, while reserving judgment about creating pre-determined categories to let the unsupervised learning provide inductive estimations about the types of bias. Event codes and their associated Goldstein scores are discrete; this discrete ontology gives little space for nuance. Rather than trying to have our learning algorithm classify articles according to CAMEO, we instead propose an algorithm that scores articles on a continuous –10 to 10 conflict to cooperation scale (Goldstein, 1992; Schrodt et al., 2008). This allows two cultures to agree that an event occurred and even on the type of the event while still holding different interpretations of events, allowing nuance to appear.

5 Moving Forward: Language and Culture through a Kaleidoscope

The larger goal of de-biasing event data for social science research is to functionally improve our capability to identify foreign policy blind spots, and avoid costly misinterpretations and misunderstandings (Downes, 2017). In international politics, biases in AI can generate erroneous and inaccurate forecasting models that miss critical events such as the Arab

Spring, or get the direction or magnitude of predictions wrong (Wang et al., 2016). Event data is a particular genre of political data that reports and encodes the actions and relationships between actors in the international system, including countries, NGOs, individuals, and groups of people. Event data sets (Lautenschlager et al., 2015) represent a significant conceptual, technological, and financial investment and are used to inform government policy decisions, but algorithms ignore temporal and linguistic nuances that bias event code generation and political forecasting models (Bagozzi et al., 2018; Geddes, 1990; Wang et al., 2016).

To answer the following questions about cultural and linguistic bias in event data, we need a multilingual, multi-cultural "kaleidoscope" algorithm (see Figure 3.3) to translate between cultures: (1) To what extent do world media report on similar events differently? (2) What semantic properties underlie the differences in how media describe the same or similar categories of events? (3) What event data categories may be missing or misclassified given the constraints of the Goldstein coding scheme (Goldstein, 1992)? (4) If we apply the word embeddings of one culture to the news corpus of another culture, how do event codes change? In answering these questions, we hypothesize that we can quantify the linguistic distances of meaning between news reports from various world cultures, and second, by projecting semantic spaces of one culture onto another using transfer learning (Shin et al., 2016), generate cross-cultural interpretations of messages and events. An event data kaleidoscope would facilitate seeing news through different cultural and linguistic lenses, where shifting the language produces a change in perspective.

Figure 3.3 Language and culture through a kaleidoscope.

The goal of the kaleidoscope process is to generate a computational model that allows researchers to explore how different languages and cultures perceive events. A simplified model of the kaleidoscope process works as follows: the source cultural context and language is chosen from the list of world cultures, and a set of news stories is selected. The description of events is assigned event data codes using word embeddings from that particular cultural perspective. These should reflect the events and priorities of the target sociolinguistic perspective being modeled. The next iteration of the process is to use word embeddings from a different sociolinguistic perspective, and then to compare the codes generated using that framework. If the cultures use different words, or use words differently, to describe the process, then those variations can be captured both in the absolute differences between the numerical word vectors, and also in the outcome codes generated by the event data encoding process.

The essence of cultural difference, and cultural bias, lies in capturing how societies perceive and encode events in language, and how those expressions are in turn encoded in event schema. Different cultural interpretations of events could potentially be modeled by switching out the standard word embedding in the source language, for one of our culturally cognizant word embeddings with the kaleidoscope process. Event data presently provides only one cultural lens through which to see the world. Failing to understand varying cultural interpretations of events and messages can lead to foreign policy blind spots that may waste finite resources and cost lives. A project following the principles detailed here will generate a cultural translation tool using a kaleidoscope algorithm to "translate" between interpretations of news media reports of international political activities. It would be able to measure the semantic differences in perception between countries and cultures (Antoniak & Mimno, 2018), to quantify cultural interpretations of messages, events, and event categories.

In this sense, the bias "bug" is leveraged as a "feature" for understanding not only how international actors conceptualize events, but also how recommender and decision-support systems are used by both the public for general information consumption and by public and private sector analysts to inform policy-making and business decisions. Artificial intelligence makes new insights into the nuances of event data coding that help researchers overcome previous methodological limitations, such as conceptual simplicity and data cleanliness. In trying to model real world scenarios, researchers must toggle between parsimony and complexity. Each has its own merits and limitations. AI can be a bridge between these competing demands—of descriptiveness and succinctness—to better represent the nuanced details of international events, with more precision in the empirical models that lead to better forecasting and more accurate representations of conflict and cooperation dynamics in the world.

Note

1 See: WEIS/KEDS (Gerner et al., 1994; Schrodt & Hall, 2006), IDEA/PANDA (Bond et al., 1994), SHERFACS (Sherman, 1994), SPEED and Phoenix (Althaus et al., 2017; Nardulli et al., 2019), CAMEO (Gerner et al., 2002; Goldstein, 1992), GDELT (Leetaru, 2014), EMBERS AutoGSR (Saraf & Ramakrishnan, 2016), and ICEWS (Lautenschlager et al., 2015).

References

Althaus, S. et al. (2017). *Cline Center historical phoenix event data v.1.0.0*. http://www.clinecenter.illinois.edu/data/event/phoenix/.

Antoniak, M., & Mimno, D. (2018). Evaluating the stability of embedding-based word similarities. *Transactions of the Association for Computational Linguistics, 6*, 107–119.

Bagozzi, et al. (2019). The prevalence and severity of underreporting bias in machine-and –human-coded data. *Political Science Research and Methods, 7*(3), 641–649.

Bolukbasi, T. et al. (2016). Man is to computer programmer as woman is to homemaker? Debiasing word embeddings. In *Advances in Neural Information Processing Systems*, 4349–4357.

Bond, D., Bennet, B., & Voegele, W. B. (1994). Panda: Interaction events data development using automated human coding. In *Proceedings of the Annual Meeting of the International Studies Association in Washington, DC*.

Bueno de Mesquita, B. (2010). *The Predictioneer's game: Using the logic of Brazen Self-Interest to see and shape the future*. Random House Trade Paperbacks.

Butler, C. K., & Gates, S. (2012). African range wars: Climate, conflict, and property rights. *Journal of Peace Research, 49*(1), 23–34.

Caliskan, A., Bryson, J. J., & Narayanan, A. (2017). Semantics derived automatically from language corpora contain human-like biases. *Science, 356* (6334), 183–186.

Cha, V. D. (2002). North Korea's weapons of mass destruction: Badges, shields, or swords? *Political Science Quarterly, 117*(2), 209–230.

Churchland, P. S., & Sejnowski, T. J. (1994). *The computational brain*. MIT Press.

Cibelli, E. et al. (2016). The Sapir-Whorf hypothesis and probabilistic inference: Evidence from the domain of color. *PloS One, 11* (7), e0158725.

Downes, C. (2017). Strategic blind-spots on cyber threats, vectors and campaigns. *Cyber Defense Review*. http://cyberdefensereview.army.mil/.

Druckman, J. N., & Lupia, A. (2000). Preference formation. *Annual Review of Political Science, 3*(1), 1–24.

Eck, K. (2012). In data we trust? A comparison of UCDP GED and ACLED conflict events datasets. *Cooperation and Conflict, 47*(1),124–141.

Finch, A., Hwang, Y. S., & Sumita, E. (2005). Using machine translation evaluation techniques to determine sentence-level semantic equivalence. In *Proceedings of the Third International Workshop on Paraphrasing* (pp. 17–24), IWP.

Ganor, B. (2002). Defining terrorism. *Media Asia, 29*(3), 123–133.

Geddes, B. (1990). How the cases you choose affect the answers you get: Selection bias in comparative politics. *Political Analysis, 2*, 131–150.

Gerner, D. J., Schrodt, P A., Francisco, R. A., & Weddle, J. L. (1994). Machine coding of event data using regional and international sources. *International Studies Quarterly, 38*(1), 91–119.

Gerner, D. J., Schrodt, P. A., Yilmaz, O., & Abu-Jabr, R. (2002). Conflict and mediation event observations (CAMEO): A new event data framework for the analysis of foreign policy interactions. *International Studies Association, New Orleans*. https://d1wqtxts1xzle7.cloudfront.net/49779845/Conflict_and_Mediation_Event_Observation20161021-31209-kh066 2-with-cover-page-v2.pdf?Expires=1647644850&Signature=A1IFEXSda q60g4Co6heuC4eWast54Rq5IaKszmQSl6Xcjcc62ReqGnRbqnEs69 xoMZ99o~O3WDMrD6mXM4wwypVWwfBJ9D9vuXPF9cm~IM L94H26YgrHc2ohbVKYo1Au4zmKnhm7wqAxZ-zFcYLYdYiOM8 pdN6GoyC7wPmuVPMj46TGuxqjSa1gQ99ni1Vo6dm4rgu1r-SSQ4xw ln580Mhs00jgi4cn7OlCu8koIUbgaDnrA7~p5xjYpXiPYJ9tgnL3Vk oQxxy5mlhJiNJo~VeY2V7XAIrzZDwIZKpT2FeQ-IElvkxWEopdJIaY~ fPez4xRhbpUH0YZzx3-cEuDODw__&Key-Pair-Id=APKAJLOHF5GGS LRBV4ZA.

Goldstein, J. S. (1992). A conflict-cooperation scale for WEIS events data. *Journal of Conflict Resolution, 36*(2), 369–385.

Greenwald, A. G., & Hamilton Krieger, L. (2006). Implicit bias: Scientific foundations. *California Law Review, 94*(4), 945–967.

Gumperz, J. J. (1968). The speech community CO. *Linguistic Anthropology: A Reader*, 43–52.

Labzina, E., & Nieman, M. D. (2017). State-controlled media and foreign policy: Analyzing Russian-language news. *European Political Science Association, Milan, Italy*.

Lasswell, H. D. (1948). The Structure and function of communication in society. *The Communication of Ideas, 37*, 215–228.

Lautenschlager, J., Shellman, S., & Ward, M. (2015). ICEWS event aggregations. *Harvard Dataverse, V3*. https://doi.org/10.7910/DVN/28117,

Leavy, S. (2018). Gender bias in artificial intelligence: The need for diversity and gender theory in machine learning. In *Proceedings of the 1st International Workshop on Gender Equality in Software Engineering*, 14–16.

Leetaru, K. (2014). *GDELT project*. https://www.gdeltproject.org/.

Lucy, J. A. (2015). Sapir-Whorf hypothesis. In J. D. Wright (ed.), *International encyclopedia of the social &behavioral sciences* (Second Edition), pp. 903–906. Elsevier. https://doi.org/10.1016/B978-0-08-097086-8.52017-0.

Mallery, J. C. (1994). Beyond correlation: Bringing artificial intelligence to events data. *International Interactions, 20*(1–2), 101–145.

Mikolov, T., Chen, K., Corrado, G., & Dean, J. (2013). Efficient estimation of word representations in vector space. *arXiv:1301.3781*. https://arxiv.org/abs/1301.3781.

Mikolov, T., Yih, W., & Zweig, G. (2013). Linguistic regularities in continuous space word representations. In *Proceedings of the 2013 Conference of the North American Chapter of the Association for Computational Linguistics: Human Language Technologies*, 746–751.

Morrison, T., & Conaway, W. A. (2006). *Kiss, bow, or shake hands: The bestselling guide to doing business in more than 60 countries*. Adams Media.

Nardulli, P. et al. (2019). Social Political Economic Event Dataset (SPEED), In P. Liberia & L. Sierra (eds.). (1979–2008). Cline Center for Advanced Social Research. V1.0.0. August 29. University of Illinois at Urbana-Champaign. https://doi.org/10.13012/B2IDB-7407320_V1.

Nieman, M. D., Chyzh, O., & Gibler, D. M. (2018a). *Modeling structural selection in disaggregated event data*. Iowa State University Digital Repository.

Nieman, M. D., Chyzh, O., & Gibler, D. M. (2018b). *Structural selection for micro-level events data*. Iowa State University Digital Repository.

Norris, C. (2016). Petrarch 2: Petrarcher. arXiv:1602.07236.

NPR. (2016, March 14). Can Computers Be Racist? The Human-Like Bias of Algorithms. NPR.Org. https://www.npr.org/2016/03/14/470427605/can-computers-be-racist-the-human-like-bias-of-algorithms.

Osorio, J. et al. (2019). Translating CAMEO verbs for automated coding of event data. *International Interactions*, 45(6), 1049–1064.

Pennebaker, J. W., Boyd, R., L., Jordan, K., & Blackburn, K. (2015). The development and psychometric properties of LIWC2015. *UT Faculty/Researcher Works*. https://utexas-ir.tdl.org/handle/2152/31333.

Pennington, J., Socher, R., & Manning, C. (2014). Glove: Global vectors for word representation. In *Proceedings of the 2014 Conference on Empirical Methods in Natural Language Processing (EMNLP)*, 1532–1543.

Saraf, P., & Ramakrishnan, N. (2016a). EMBERS Autogsr: Automated coding of civil unrest events. In *Proceedings of the 22nd ACM SIGKDD International Conference on Knowledge Discovery and Data Mining*, ACM, 599–608.

Saraf, P., Self, N., & Ramakrishnan, N. (2016b). Who, when, where and why? Visualizing civil unrest events. In *Proceedings of the IEEE VIS 2016 Workshop on Temporal Sequential Event Analysis*.

Schrodt, P. A. (2012). Precedents, progress, and prospects in political event data. *International Interactions*, 38(4), 546–569.

Schrodt, P. A. (2013). *CAMEO Scale: Version 0.5B1*. http://eventdata.parusanalytics.com/cameo.dir/CAMEO.SCALE.txt.

Schrodt, P. A. (2015). *Comparison metrics for large scale political event data sets*. Paper presented at the New Directions in Text as Data, New York University, 16–17 October 2015.

Schrodt, P. A. (2017). A Practical guide to current developments in event data. *International Methods Colloquium, Rice University Webinar*.https://www.methods-colloquium.com/.

Schrodt, P. A., & Hall, B. (2006). Twenty years of the Kansas event data system project. *The Political Methodologist*, 14(1), 2–8.

Schrodt, P. A., Yilmaz, O., Gerner, D. J., & Hermreck, D. (2008). The CAMEO (Conflict and Mediation Event Observations) actor coding framework. In *2008 Annual Meeting of the International Studies Association*.

Sherman, F. L. (1994). Sherfacs: A Cross-paradigm, hierarchical and contextually sensitive conflict management data set. *International Interactions*, 20(1–2), 79–100.

Shin, H. C. et al. (2016). Deep convolutional neural networks for computer-aided detection: CNN architectures, dataset characteristics and transfer learning. *IEEE Transactions on Medical Imaging, 35*(5), 1285–1298.

Si, Y., & Reiter, J. P. (2013). Nonparametric Bayesian multiple imputation for incomplete categorical variables in large-scale assessment surveys. *Journal of Educational and Behavioral Statistics, 38*(5), 499–521.

Slobin, D. I. (1996). From 'thought and language' to 'thinking for speaking.' In J. J. Gumperz & S. C. Levinson (eds.), *Rethinking linguistic relativity*, pp. 70–96. Cambridge University Press. (Reprinted in modified form from *Pragmatics*, 1, 1991, pp. 7–26).

Solaimani, M. et al. (2016). Spark-based political event coding. In *2016 IEEE Second International Conference on Big Data Computing Service and Applications (BigDataService)*, 14–23.

Solaimani, M. et al. (2017). RePAIR: Recommend political actors in real-time from news websites. In *2017 IEEE International Conference on Big Data (Big Data)*, 1333–1340.

Solaimani, M. et al. (2017). APART: Automatic political actor recommendation in real-time. In *Social, Cultural, and Behavioral Modeling: Lecture Notes in Computer Science*, Springer, Cham, pp. 342–348.https://link.springer.com/chapter/10.1007/978[[sbn]]3[[sbn]]319[[sbn]]60240[[sbn]]0_42.

Staff of PLoS one. (2016). Correction: The Sapir-Whorf Hypothesis and probabilistic inference: Evidence from the domain of color. *PloS one, 11*(8), e0161521.

Sydell, L. (2016). Can computers be racist? The human-like bias of algorithms. *National Public Radio*.https://www.npr.org/2016/03/14/470427605/can-computers-be-racist-the-human-like-bias-of-algorithms.

Tseng, C., Carstensen, A. B., Regier, T., & Xu, Y. (2016). A computational investigation of the Sapir-Whorf hypothesis: The case of spatial relations. In *Proceedings of the 38th Annual Meeting of the Cognitive Science Society*, pp. 2231–2236.

van Atteveldt, W., Kleinnijenhuis, J., & Ruigrok, N. (2008). Parsing, semantic networks, and political authority using syntactic analysis to extract semantic relations from Dutch newspaper articles. *Political Analysis, 16*(4), 428–446.

Wang, W., Kennedy, R., Lazer, D., & Ramakrishnan, N. (2016). Growing pains for global monitoring of societal events. *Science, 353*(6307), 1502–1503.

Ward, M. D. et al. (2013). Comparing GDELT and ICEWS event data. *Analysis, 21*(1), 267–297.

Weidmann, N. B. (2016). A closer look at reporting bias in conflict event data. *American Journal of Political Science, 60*(1), 206–218.

Whorf, B. L. (1940). *Science and linguistics*. Bobbs-Merrill.

Wu, X. et al. (2019). On constructing a knowledge base of Chinese criminal cases. *arXiv:1910.07494 [cs]*. http://arxiv.org/abs/1910.07494.

Zou, J, & Schiebinger, L. (2018). AI can be sexist and racist—It's time to make it fair. *Nature, 559*(7714), 324–326.

Part 2
AI in the Classroom

Chapter 4

Truth in Our Ideas Means the Power to Work

Implications of the Intermediary of Information Technology in the Classroom

COL James Ness, LTC Lolita Burrell and David Frey

> Wovon Man nicht sprechen kann, darüber muß Man schweigen
> (About which one cannot speak, one must be silent)
> (Wittgenstein, 1921/2010)

1 Introduction

In 1913, Thomas Edison predicted that every branch of knowledge would be taught using motion pictures, portending the demise of books (The Economist, 2013). Although film is a powerful media often leveraged in the classroom, written communicative means remain the instructional foundation for education. With the current proliferation of technology in classrooms, the utility of books, teachers, and the classroom itself are again being questioned. However, like the advent of motion pictures, the supplanting of teachers and classrooms by information technology may be overdone (Paulson, 2014; Wittgenstein, 1921/2010). Nevertheless, advances in information technology have had a profound effect on education. The advance is two-fold. First, analogous to the Great Library of Alexandria, the wealth of information afforded through technology provides a means to discover and learn an abundance of concepts and skills (Mitra & Dangwal, 2010). However, to benefit from that wealth of information, one must have an affinity for the material, a certain deftness with technology, and exercise self-regulation to remain deliberate in the learning process (Broadbent, 2017; Pintrich, 2000). Second, the Internet and other virtual technologies afford interactive platforms where, in real time, the teacher can meet with students and where groups can collaborate. In real time interactions, the teacher or peer collaborators can actively guide participation in the content to advance understanding of the material. The interactions are, however, indirect as they are mediated through the technology.

DOI: 10.4324/9781003030928-7

Access to technology's wealth of information is achieved through its maintenance and facilitation functions. The Internet, for example, provides access to vast amounts of information. Abstraction of that information is facilitated through powerful algorithms designed to search, retrieve, and organize information across database structures. Thus, technology afforded information, as is inherent to a system based in logic, are deterministic. As such, technology captures the deterministic aspects of cultures' literacy and numeracy. However, there does exist, within communicative systems, an indeterminism. That indeterminism emerges in the activity of establishing and maintaining intersubjectivity between the individual and the object of study, between the student and teacher, and within collaborative groups (Piaget, 1978; Vygotsky, 1988). This indeterminism presents challenges to technologically mediated activities, challenges this chapter will explore.

Before one can assess technology in the classroom, one must first assess the nature of human cognition to which technology is to be applied. The assessment to follow is pragmatic in the sense of William James. James (1906–1907) advocated that all theories become instruments, in other words, useful tools, and not answers to enigmas in and of themselves. Although less than a law of nature, theory does provide a systematic framework to discover the nature of things through its utility. Cognitive theories, as instruments, specify practice toward producing optimal outcomes. The inferences concerning the relationship between the practice and outcome are drawn to inform the understanding of the nature of human knowledge. Thus, the presentation of theory in this chapter will focus on the practical consequences of applying theory to develop best educational practices. Because education involves a culture's literacy and numeracy, we begin with a discussion of language as the communicative medium to maintain, facilitate, and induce individual and societal knowledge.

2 The Indeterministic Nature of Language

Wovon Man nicht sprechen kann, darüber muß Man schweigen
(About which one cannot speak, one must be silent)
(Wittgenstein, 1921/2010)

Wittgenstein's musings on ethics and metaphysics point to a potential limit to language, in that logic (e.g., semantics) often comes prior to experience (e.g., pragmatics) and therein lies a caution. The logic of language, he suggested, is without sense (i.e., perception), although it is not nonsense. Therefore, it is hard to express that which is sense-full without experience. Both the sense-full and the logical are conveyed in communicative systems through the exchange of signs and symbols. Specifically, communicative systems mean any activity in the exchange and development of signs. The development is a continual cultural process called the semiotic, which Peirce argues to be comprised of a triadic of the object, its sign, and the interpretant (Atkin, 2013). The interpretant is where the meaning is made manifest to the individual through their perception of the conveyed sign.

The logical aspects of such systems are generally deterministic as words have meanings determined in authoritative cultural indices such as dictionaries. Standard rules of grammar for a given language determine structures of syntax, which are also generally deterministic. The logical or semantic component is a culture's formal assignment of signs to an object, which renders a literacy and numeracy scaffold for communicative systems. The influences of the sense-full on the logic of the semantic can be seen in etymology and in numeracy in the evolution of numerical systems.

Upon the literacy and numeracy scaffold is the sense-full aspect inherent in the interpretant. The sense-full is generally indeterministic as experiences and the redintegrations of those experiences are activities rather than extant entities, and thus not known *a priori*. Within the sense-full, upon the scaffold of the semantic, intersubjectivity is developed between communicative partners (Vygotsky, 1988), and between concepts and the understanding minds operating on those concepts (Piaget, 1978). In the sense-full, the indeterminate or pragmatics of language is conveyed whereas in the logic, which lacks a sensing component, the determinate or semantics of the language is conveyed (Recanati, 2002).

For example, Wasdyke and Ness (2019) employed a Thurstone method (Thurstone, 1928) to evaluate the lexicon of mood across event groups in the Army West Point Women's track team (Wasdyke & Ness, 2019). Across all event groups, there were significantly fewer words associated with sad moods compared to happy and neutral moods. Between event groups, there were differences in word meaning, particularly those words and phrases associated with the expression of sad moods. For example, long distance runners associated "reading more books" with sad moods versus the women in field events, particularly throwing events, associated "eating more" with sad moods. In the semantic of "sad mood", disturbances in eating are implicated, but changes in reading habits are not (International Classification of Diseases, 10th Edition, 2020) and thus not understood as sad outside the long-distance runners' social niche. The context for reported sad moods were sidelining injuries that separated the athletes from the team, implicating a regulatory role of the social niche as described by Hofer (1984) and in the accounts of those raised in institutions (Greene, 2020). The data speak to the indeterminacy or pragmatics of the communicative system with shared experience defining the meaning and the social niche having a regulatory influence on mood states.

Language, like all things living, has both an evolutionary and an ontological history. Languages evolve as systems of written and oral signs to convey a semiotic such that there is an intersubjectivity among communicative minds. As Peirce argued, the semiotic includes the object, its sign, and the interpretant where the meaning of the sign is made manifest in the communicative mind. The interpretant is the primary agent in establishing and maintaining intersubjectivity. This is evident in the human tradition for diffusing ideas—that of storytelling. Representations of experiences, such as cave paintings, are the oldest traditions of recounting events through storytelling, imparting narrative, lessons, and projecting affect. Storytelling structures information in part–whole relations affording the viewer/listener/reader schematic frameworks to interpret past, present, and future analogous events (Campbell, 1972; Mandler, 1979). The analogies induced in the reader's or listener's mind are a form of *poiesis*, that of bringing something into being that did not exist before in the communicative minds (Lord, 1971). Thus, language is an activity between communicative partners that serves the functions of inducing new knowledge, facilitating remembering of existing knowledge, and maintaining a record of knowledge.

Of these functions, technology performs maintenance and facilitation functions well, as they are deterministic. Technology can expeditiously process vast amounts of information and is thus a powerful tool for accessing and retrieving data maintained in electronic libraries (Loh, 2018). Induction requires the indeterminism of experience and thus a human to make sense of the information within their place and time. Technology has isomorphisms for indeterminism such as Monte Carlo algorithms, but as algorithms, they are, by nature, deterministic and cannot move outside the logical loop of the algorithm. Induction of new knowledge as proposed by Popper (1966) results from breaking loops of determinism to see new possibilities, which is something technology, as a determinant system, is unlikely ever to achieve.

3 The Construct of Mind and the Implications of Intersubjectivity

If the educator is replaced by an artificial intelligent system or other technology, can new knowledge be created? The question requires a discussion of what we mean by the construct of mind, which is, as Whitehead (1919) argues, a psychic addition to nature and not nature in and of itself (Table 4.1).

To avoid confounding theoretical perspectives and associated presuppositions, a "levels of analysis" approach is often taken such

Table 4.1 Presuppositions of the mechanistic and organismic world views

	World View	
Theme	Mechanistic	Organismic
Metaphor	Machine: ultimately reducible to constituent parts	Organism: more than the sum of constituent parts
Causation	Behavior: Described in terms of discrete chains of cause and effects	Cause of developmental change inherent in the nature of the organism
Organism	Reactive/Passive	Active
Developmental change	Quantitative	Qualitative
Timing	Age dependent	Age independent
Epistemology	Naive realism	Constructivism
Interests	Behavior	Organization/change
Stimuli	Causal	Facilitate, maintain, and induce
Teleology	No goal or end	Regulative sense/adaptive
Evaluation	Outcomes	Processes
Continuity of Change	Continuous, gradual, cumulative	Discontinuous, rapid, stage-like
Direction of Change	Bidirectional	Unidirectional

that generalizations from the theory are congruent with the context from which it is derived. Table 4.1 summarizes the presuppositions of each worldview as derived from Brainerd (1978). In general, at lower levels, the mechanistic worldview is more robust as structure is a strong influencer of function. As one moves up through the levels of organization from neurological to social, the structure/function relationships become more indeterminate and organismic models appear more robust.

The construct of mind is a construct of many integrated levels, some of which are explained by the activities of other levels and some are unique to a level (Hofstadter, 1979). Those levels closer to nature such as mechanisms for flow of ions across axonal membranes are well explained. Those levels, in which psychic additions are made, such as cognitive, are more enigmatic. Espoused in this chapter is a construct of mind as more than the organ of the brain but is never without a brain.

Mind is made manifest by the functions of the brain and thus a levels of analysis approach. More formally, mind, as Bateson (1979) argues, although an agent, should not be treated as having concrete or material existence, but should be considered as a process with multiple levels that interact with each other. Among these levels are the concrete organ called the brain, the states of the whole body, that which is discoverable in the world, and the evolving semiotic of the communicative exchange, all of which serve to construct thought and language.

The definition of mind does not eschew mechanistic metaphors of the mind as they do speak to limits of signal processing and structural organization within the nervous system, which are a component of mind. The mechanistic derived models, particularly the information processing models, have their roots in Atkinson and Shiffrin (1971), who derived their model from Shannon's (1948) Information Theory. Information Theory informs those aspects of the semiotic involving the process of converting the signals from outside the body and transmitting the signals along neuronal pathways. The transmissions are essentially analogue as these neuronal impulses transmit duration, intensity, and rate information. These signals do have limits as neurons have refractory periods, classes of neurons have different cable properties, and there are limits to transducing agents such as opsins, all of which yield generally logarithmic functions in making comparative judgements of sensations (Burt, 1960; Stevens, 1957). These limits are also shown in Fitts' Index of Performance of various neuromuscular channels in that there are optimums for channel performance, vice speed/accuracy trade-offs (Fitts, 1954).

Fitts' Index of Performance likely reflects the somatotopic organization of the motor and premotor cortex, of which the prefrontal cortex has no direct connections to the motor nuclei (Goldman-Rakic, 1996). This likely explains the issue of speed/accuracy trade-off, in that when the prefrontal cortex is engaged, speed accuracy trade-offs are manifest as response initiations and inhibitions involve the prefrontal cortex (Goldman-Rakic, 1987). If the behaviors are maintained within the neuromuscular channel, then the Index of Performance predicts behavior. If the behaviors are influenced by the premotor cortex, then speed accuracy trade-offs are manifest when the conscious act of initiating and inhibiting motor behavior drives responses outside neuromuscular channel limits.

Limits are also seen in the extension of the mechanistic models in defining memory. These models have shown limits of capacity of the hypothetical constructs of iconic, short-term, and working memory (Baddeley, 2000). Nonetheless, these models maintain that long-term memory is seemingly limitless. Arguments for limits of the working memory system, although based in extensions of information theory, have shown nominal application informing design of human operated systems (Seow, 2005). The reason may be that limits are made manifest under conditions that do not generalize well to ecologically valid contexts (Goldberg, 2009). Thus, the short-term/long-term memory dichotomy may be, in part, an artifact of experimental conditions rather than an explanation of the memory phenomenon. Notwithstanding, the mechanistic derived models of memory do speak to the logical underpinnings of the electrical and chemical processing properties of the brain. The brain certainly remains an enigma with its various levels of activities in response to transduced external energies. Interestingly, signaling patterns feeding forward from primary sensory cortices to the prefrontal cortex preserve spatial and temporal patterns in the redintegrations (Goldman-Rakic, 1996). The underpinning neurochemical and electrical activities reflect Wittgenstein's concepts, as these activities are without sense but are not nonsense. With increasing levels of organization come increasing contributions of senseful experience and thus less of a deterministic relationship between structure and function.

Memory, as operationalized from a mechanistic set of presuppositions is something one has, which tends to suggest storage limits and a short-term / long-term memory dichotomy. Memory as operationalized from an organismic set of presuppositions is something one does, and thus limited more by experiential than structurally related memory storage constraints. Under mechanistic derived conditions, as in Ebbinghaus-like memory drum tasks (Haupt, 2001) where the experimenter determines what is to be remembered, when it is to be remembered, and how it is to be remembered, limits are reliably evident in the hypothetical system—memory. Under conditions where, as argued by Goldberg (2018), the person engaged in the memory task is allowed to decide what to remember, when to remember it, and how to remember it, the construct of working memory becomes the process of selecting task relevant information in the course of behaving rather than encoding and retrieving to and from a hypothetical set of stores. Thus, the dichotomy of having short and long-term memory stores is replaced by the concept of memory as something the person does, and that activity is best described as salience assignment.

From an organismic set of presuppositions, memory and cognition become behaviors arising from a transaction between biological activities and information afforded in the world. This implies memory is both a central as well as an external process. It is central in that salience is assigned and refined with experience and with the accompanying redintegrations prompted by the experience. It is external in that affordances in the effective environment are discoverable and constitute a knowable world (Norman, 2013). Thus, under ecologically generalizable conditions, limits seem to be more a function of experience that yield abilities to see contingencies (signs) among the objects in the discoverable world than limits in hypothetical working memory stores.

Skinner makes the contingency case for operant behavior by suggesting that what guides behavior is an ability to respond to contingencies of reinforcement in the environment. What develops then is the ability to perceive ever more complex contingencies in the environment acquired through sense-full experiences of life (Skinner, 1972). The operant idea certainly acknowledges the affordances in the world but leaves unanswered the activities of the "intervening variable" of mind in the discovering, knowing, and remembering what is afforded in the knowable world.

Within the mind, the saliency of certain signals in certain contexts is assigned relevance while others are not. The idea of context dependent salience assignment is an idea posited by Goldberg (2018) in the definition of memory as salience assignment to signals while functioning in a niche. As salience assignment, memory and cognition are not things in the brain or things in the world but are a phenomenon of activity of the person negotiating a niche and as such a result of the sense-full nature of knowledge acquisition.

Within the brain, Zal'mason (1926) directly observed the effects of experience on the induction of organizational change in physiological functioning through the discovery of the existence of "temporal dominant foci". These "foci", as described by Pribram (1991), can be considered pacemakers in that they appear to be a source of organization for whole brain activity in the presence of particular stimulation. Specifically, Zal'mason (1926) conditioned a dog to raise its right leg to the sound of a tone. After this conditioned response was well established, its right motor cortex (which controls the left side of the body) was exposed. During the performance of the conditioned reaction, a patty of strychninized filter paper was placed on the exposed right motor cortex in order to chemically excite it. Immediately the dog switched the responsive leg raising the left forepaw to the conditioned stimulus. The focus of neural activity, which had been established through conditioning and which dominated the functions of the motor cortex, was now overshadowed by a new "temporary dominant focus" established in the dog's brain by the chemical excitation from the strychnine. Thus, the existence of these foci was shown to be not exclusively dependent on endogenous mechanisms, but require, for their existence, reliable input from the stimulus environment. These data not only force us to view the environment as a reliable, dependable, and predictable contributor to the development of memories, but also the environment, in the role of reliable force, can induce changes in brain functioning. Moreover, these changes in functioning can lead to structural change in the form of changes in the number and organization of dendritic projections (Lynch, 1971).

The influence of experience in the development of human knowledge occurs at least at two levels. The levels are defined by the source of intersubjectivity between the knowing person and the knowable world. One is intrapersonal, between the individual's schemes and what they afford in the knowable world. The organizing structures of the mind are developed to better adapt to and assign salience in interactions with the discoverable world. The other is interpersonal between communicating minds sharing and refining the salient features of experiences yielding a vicarious experience to the inexperienced and revealing alternatives to future similar experiences to the experienced. In both cases, it is the sense-full that induces new understanding and knowledge.

4 Relationships between Thought and Language

The realization that experience can induce organizational change is a major philosophical break from the traditional understanding of the relationship between endogenous and exogenous influences on physical and behavioral development. Instead of suggesting that physical and behavioral traits are predetermined by structural complement, the suggestion here is that epigenetic relationships are driving physical and behavioral development. However, this is not the kind of epigenetic relationship that Waddington (1966) described by way of his epigenetic landscape. What is being argued here is that, with increased organization the epigenetic landscape itself changes, thus providing for greater variability in behavior instead of greater canalization as is the case in the more deterministic characteristics of physical and physiologic.

Greater variability in the activity of memory comes as one moves levels of analysis from the individual learner where intersubjectivity is within the individual in the cognitive development described by Piaget (1978), to communicative groups where the locus of intersubjectivity is between communicating individuals in the cognitive development described by Vygotsky (1988). For Piaget, learning occurs through an individual's active engagements with the environment (Table 4.2). Major advancements in the individual's thinking are a result of an equilibration of schemes used to discover the world to that which is discoverable in nature. The equilibrations are internal to the hypothetical mental structures and thus the intersubjectivity is within the individual. For Piaget, thought precedes language as schemes (thought) must be developed to interpret the concrete world as an abstract (language). Communication is best between equal learning partners as schemes and thus thought and its resultant language are relatively congruent.

For Vygotsky, language precedes thought because literacy is the scaffold for establishing intersubjectivity between individuals (Table 4.2). The learning takes place between unequal learning partners and occurs as a result of functioning on the edge of one's competence in a discoverable world. Because language precedes thought, thought becomes a social phenomenon and not something exclusive to the mind of the individual. In support of the idea that language precedes thought, Vygotsky demonstrates that societies maintain means to facilitate intersubjectivity

Table 4.2 Compare and contrast of Vygotsky and Piaget

A. Vygotsky	B. Piaget
C. Instructional technologies	D. Operating on mental structures
E. Focus: Social basis of mind	F. Focus: Individual as starting point
G. Locus of intersubjectivity is between partners	Locus of intersubjectivity is between the individual's schemes and what they afford.
H. Guided participation: Zone of proximal development between unequal partners	Cognitive schemes guide behavior to assimilate information and accommodate behaviors for further assimilation of information.
Intra- to interpersonal plane of cultural tools	Qualitative shift in perspective
Cognitive development is historico-cultural internalization from social to individual	Cognitive development occurs with equilibration of schemes with experience.
Social Activity is the unit of analysis	Individual is the unit of analysis
Inequality of learning partners	Equality of learning partners at concrete operations (≈age 7)
Zone of proximal development	
Language precedes thought	Thought precedes language

between and among individuals. Those means are a culture's literacy and numeracy tools, which are taught in schools and conveyed through technology. Additionally, cross-cultural exchange adds another layer to the social basis of the mind and to resulting knowledge production (Campbell, 1972).

5 Implications for Technology in the Classroom

Within the context of what is understood concerning the human nature of knowledge and its deterministic and indeterministic components, the question concerning technology's role in the classroom can be assessed. At the limits of the question is (Wittgenstein, 1921/2010) technology replacing the educator, leaving only learner and machine interactions and (The Economist, 2013) a classroom devoid of technology and thus no guided participation in a tool pervasive in society. Inherent in the question of technology in the classroom is the ethical standard of beneficence.

An educational experience devoid of technology does not meet a beneficence standard, as a curriculum devoid of technology does do harm. The purpose of education is to ensure the acquisition of specialized knowledge and skills central to a functioning society (Rogoff, 1991). Technology is an advancement of societies' literacy and numeracy, which are skills central to the functioning of society. Thus, a curriculum devoid of technology meets neither the goals of education nor the needs of society and therefore does harm as the learner will be ill-equipped to function in society.

At the other limit, a classroom in which technology replaces the teacher would, due to the current nature of technology, be a deterministic logical system. Although its determinism is part of human knowledge acquisition, logic is devoid of the indeterministic quality of human knowledge acquisition, that of experience. Such a logically based educational system would promote thinking, as one is inclined to think (Peirce, 1887). There would be a tendency toward homogeneity and elimination of doubt as the authority for knowledge would be relegated to the technology (Turgue, 2012). Further, the emotional regulatory aspect of human-to-human social interchange, a critical aspect for regulation of behavior, would be altered.

To understand the effect of interacting with technology on emotional regulation, Lucas et al. (2014) compared responses of volunteers interacting in human versus computer framed communicative exchanges where the volunteer reported on their psychological well being. The aim of the study was to determine if, in a mental health care context, individuals would be more honest in an interaction they believed to be with an intelligent agent versus one believed to be with a human. In the intelligent agent condition, some volunteers interacted with an intelligent agent and others interacted with a tele-operated agent. In both conditions, the volunteers believed they were interacting with an intelligent agent. In the human interaction condition, the volunteers also interacted with either an intelligent agent or a tele-operated agent. However, in this condition, the volunteers believed they were interacting with a human. Subsequent to the interactions, volunteers self-reported fear and impression management as measured in responses to items on the Brief Fear of Negative Evaluation Scale and the Balanced Inventory of Desirable Responding. Degree of sadness expressed was measured using the Computer Expression Recognition Toolbox, which is programmed to score facial actions typically associated with emotions. In addition, an observer who was blind to the conditions scored answers to eight questions operationalized as indicators of degree of self-disclosure. When volunteers believed they were in the presence of an intelligent agent, the data indicated less fear and impression management and showed a 30% increase in intensity of displayed sadness as compared to interactions believed to be with a human. Only the main effects between the human and intelligent agent framed conditions were statistically reliable. The authors concluded that in the believed presence of a virtual agent, people are more honest with their feelings than in the believed presence of a human. They further argued for the use of virtual agents at initial wellness screening. The conclusions and recommendations are justified, if in fact the pattern of responses indicate more honest reporting and not diminished self-regulation.

Certainly Lucas et al. (2014) showed an effect of human versus computer framed interaction for people expressing feelings of sadness. The question however remains: is the increase in the expressions indicative of greater genuineness or diminished regulation of emotion? The reduction in fear and impression management in conjunction with an increase in expressions of sadness can be interpreted as a more real and genuine expression of state of well-being. However, it may also indicate a diminished regulation brought about by an impoverished social environment (Thayer & Lane, 2009) and thus not truly reflective of the state of well-being.

Humans, like all social animals, are, to an extent, homeostatically open systems (Hofer, 1981) with regulatory effects of the social environment most prominent in periods where the person is experiencing a biobehavioral shift (Emde & Gaensbauer, 1981). These sensitive periods in development result in adaptations that determine the composite of morphological, physiological, and behavioral characteristics of subpopulations within a general population (Gottlieb, 1998; Kou, 1976; Oyama, 1985). For example, personality is often described as a set of traits that explain stability of behavior across contexts. Associated with personality is a biobehavioral shift in identity in the transition from adolescence to young adulthood. This shift entails forming identities and restructuring social and emotional attachments from parents to peers. During this period, personality factors are particularly mutable in the presence of a strong environment (Ness, Lewis, & Brazil, 2011) and disorganized in environments that present minimal, culturally relevant social support (Alarcón, 2009; Payer, 1996).

Concerning the latter, in the mid-1970s, doctors in a U.S. Army clinic started seeing a problem in their unit. Young healthy males were withdrawing from life, hiding in their rooms, and deteriorating in their performance. When seen in the clinic, these individuals often were crying, screaming that they simply could not deal with the Army anymore, and were shaking their limbs and bodies almost convulsively. Initial presentation involved complaints of limb numbness and tingling. Health care providers in the clinic came up with a usable albeit catch-all diagnosis "adjustment reaction of adult life". On further investigation, it was realized that the manifestation was almost entirely found in a population of young (average age of 18 and 19-year-olds) Puerto Rican males who had never been off their island before entering the Army and were then in their first operational assignment. In discussing this with one of the unit's Non-Commissioned Officers (NCO, e.g., sergeant; also, Puerto Rican), clinic staff were informed that this was a normal means of expression of stress on the island among this group, and that it was worse when the population group was not in contact with females (the implication was clear that Puerto Rican females simply would not put up with this "acting out", and through social influence would keep it under control). It was initially suspected, based on the prevailing medical logic, that the presentation might have been some kind of group hysteria, but as explained through one with sense-full experience, it was a culturally modified stress reaction. The clinic staff arranged for both the NCO and his wife (and a few other NCO spouses) to meet with the soldiers individually. The intervention worked, in that the problem did not recur, at least not to the knowledge of the clinic staff (personal communication, Brigade Surgeon D. Lam, M.D., 15 January 2010). The remedy was culturally relevant social support rendering a self-awareness in the soldiers presenting with the symptoms, helping the soldiers work through the issues to gain self-control and maintain self-esteem.

An essential regulatory mechanism in well-being is self-awareness (Duval & Wicklund, 1972). Self-awareness derives from ancient Greek aphorism ascribed to the Oracle of Delphi, "know thyself" and is a pervasive construct in well-being. According to Duval and Wicklund (1972), the self is itself indivisible; however, consciousness can act on the self in two ways. First, it can be focused on the task in which the self is engaged, a state referred to as the "Subjective Self". Second, consciousness can also be focused on the self, a state referred to as the "Objective Self". While in the Subjective Self state, one is focused on external events, where one is typically absorbed in the task, with a focus on achieving goals or otherwise influencing task-related outcomes. While in the Objective Self state, one is focused on the self as the object of study. In this state, attention is on, for example, state of arousal, likes and dislikes, and personal history. It is in this state that the individual self-monitors and self-reflects. Both objective and subjective states are susceptible to conscious activity with attention as the limiting resource. Self-awareness is shown to promote constructive outcomes in the areas of perspective taking, self-control, creative achievement, and self-esteem (Ness et al., 2010; Silvia & Duval, 2001; Wicklund & Gollwitzer, 1987). However, Duval and Wicklund (1972) do caution that an exaggerated attention on the self can lead to self-consciousness (Fenigstein, Scheier, & Buss, 1975).

In the context of education, the educator engages the student in problems and content with which the student is not quite able to manage independently. The educator promotes self-awareness of what needs to be learned while being sensitive to and ameliorating student's self-conscious feelings. In such a social exchange, the use of intelligent agents as intermediaries would effectively distance students from teachers, apprentices from masters, and masters from the work itself (Beane, 2019). Intermediaries such as intelligent agents are useful tools with a place in advancing human cognition, but if the intermediaries become the "doers" and de facto "deciders" then human cognition and culture will be diminished because memory and cognition are performative. They are things we do at the individual, social, and cultural levels. If we do not exercise those abilities, we will lose them, not to mention the self-awareness of our loss of those abilities. The argument is not for a dystopian rejection of technology. The argued is to define the art of the possible in the use of technology in education with learner and educator beneficence as the goal.

There are cases for the unequivocal beneficence of technology in the classroom, such as the "Hole in the Wall" project by Mitra and Dangwal (2010). The authors noted that children in the villages of India lagged behind similarly aged children with reliable access to schools. The "Hole in the Wall" project was conceived to ameliorate the disparity in learning by exploring the limits of self-organized learning. The authors placed computers accessible to children with content chosen by experts in the fields of study for the children to explore. The technology was used as a literacy/numeracy tool for children to explore concepts, whose presentation was scaffolded by the masters in the field. The technology afforded the children the opportunity to collaborate as equal learning partners to improve schemes toward understanding the content. As Piaget would predict, the children taught themselves. The children achieved scores comparable to those attending local state schools. Vygotsky's principles were also in play as the experts in the fields of study chose the content and scaffolded the material so as to neither frustrate nor bore the children. When the authors added the support of a mediator to help with self-esteem, self-regulation, and metacognitive awareness, by interjecting phrases such as, "I wish I could do that", "Please explain this in simple words to me", etc., the children performed equal to peers in privileged private schools. This research shows technology to be a powerful educational tool subject to critical caveats. The content must be structured by subject matter experts, equal learning partners must be free to actively engage the material, and a mediator must be present, at a minimum, to manage and instill self-esteem, self-regulation, and metacognitive awareness. Of note in these studies were the fact that children were together in small groups, in physical contact, around the screen such that they could touch each other and point to direct attention and facilitate intersubjectivity. Moreover, the mediator was in the room with the children sensing projected affect from the peer exchanges and moderating them. In this example of beneficence, technology played a significant role in delivering content. It augmented but did not replace the teacher or impoverish the social interchange, which Piaget and Vygotsky both posit as essential for learning, self-esteem, self-regulation, and metacognitive awareness.

Similar results to those of Mitra and Dangwal (2010) were reported in a study using robots as teacher's aides in a kindergarten class (Causo et al., 2017). The robots engaged the children with content developed by the educators in support of the children's curriculum. Providing resources required to support the technology was the main challenge. When the systems malfunctioned, trust in the automation quickly declined as frustrations rose. Nonetheless, technology as a teacher's aides was proven as a viable concept. Moreover, technology holds further promise as a Department of Education review of the use of technology in the classroom showed an increase in learning time for children in blended learning curriculum (U.S. Department of Education—Office of Planning, Evaluation, and Policy Development, 2009). Increased learning time suggests an increase in content knowledge, which expands the breadth and scope of understanding. This leads to new and more profound understanding, i.e., learning (Seigler, 1986).

6 Conclusions and Recommendations

Technology represents a powerful advancement in society's literacy and numeracy. Technology maintains a wealth of information and facilitates the retrieval and organization of that wealth of information giving classrooms unprecedented access to content. Thus, the use of technology in the classroom has clear benefits, provided that it is used ethically. Throughout the chapter, we apply two principal ethical standards to evaluate the efficacy of deploying technology in education: that of nonmaleficence (do no harm) and that of beneficence. These standards within a framework of cognitive theory lead us to conclude that evolving educational needs have produced an ethical obligation to include technology in the classroom and thus require teachers to develop a facility with technology, but that technology should not be viewed as a replacement for human interaction in learning.

Technology should be used for functions that facilitate and maintain knowledge. It provides benefits when employed to access the wealth of information in content areas as structured by educators and professionals in the field of study. Applying a pragmatic assessment of cognitive theory to the development and implementation of best educational practices also reveals the limits of the ethical obligation for technology's classroom use. Technology, as a logic-based system should not be used in roles as teachers or surrogate peers. Technology is without sense, although not nonsense and thus a powerful tool for maintaining information and facilitating its retrieval. Induction of new knowledge is sense-full as it requires experience and ever more refined intersubjectivity between the individual and the discoverable world and between learning partners experiencing and sharing experiences with the discoverable world.

Induction requires the indeterminism of sense-full experience, something the logic of a society's literacy and numeracy often cannot express.

Table 4.3 presents a pragmatic assessment of cognitive theory as a guide for incorporating technology in the classroom. The first two columns in Table 4.3 present the Berkeley Center for Teaching and Learning definition of learning (2020). Applying Berkeley's functions of learning as a pragmatic assessment of the use of technology is offered in the third column of Table 4.3. In this column, the use of technology in the classroom is linked with the processes of learning described by the Berkeley Center for Teaching and Learning, to promote learning through an effective use of technology as a literacy and numeracy tool.

Table 4.3 Incorporating technology in the classroom: a pragmatic assessment

I. Cognitive Construct	J. Definition	Technology
Active	From: Dewey, 1938; Piaget, 1978(7) Learning is a process of engaging and manipulating objects, experiences, and conversations in order to build mental models of the world.	Applications that allow for the visual capture of motor movements to build affordances transferable to the real-world to accelerate experience and advance the processes of salience assignation.
	From: Vygotsky, 1986(8) Learners build knowledge as they explore the world around them, observe and interact with phenomena, converse and engage with others, and make connections between new ideas and prior understandings.	Applications that allow for cooperative experiences with peers to discover content.
Builds on prior knowledge	From: Alexander, 1996, p. 89 Learning involves enriching, building on, and changing existing understanding, where "one's knowledge base is a scaffold that supports the construction of all future learning"	As demonstrated in Mitra & Dangwal, 2010, develop applications under the guidance of content experts such that the applications are appropriately scaffolded to the learner's level of understanding.

Occurs in a complex social environment	From: Bransford et al., 2006; Rogoff, 1998 Learning should not be limited to being examined or perceived as something that happens on an individual level. Instead, it is necessary to think of learning as a social activity involving people, the things they use, the words they speak, the cultural context they're in, and the actions they take. From: Scardamalia & Bereiter, 2006 Knowledge is built by members in the activity.	Applications that facilitate innovating as a team through posing problems from the field of study that the team could research, providing various techniques and support tools to facilitate discussion, collaboration, and affording a zone of proximal discovery as described by Norman, 2013.
Situated in an authentic context	From: Goldberg, 2018 Provides learners with the opportunity to engage with specific ideas and concepts on a need-to-know or want-to-know basis.	Applications, such as specialized search engines that allow for the exploration of topics of interest and a means to organize and catalogue information for subsequent retrieval.
Requires learners' motivation & cognitive engagement	From: Greeno, 2006; Kolodner, 2006 In particular, when learning complex ideas, considerable mental effort and persistence are necessary.	Learning games that motivate learning through a series of levels of difficulty, the purpose of which is not to learn the game and its goal boxes but to come away from the experience with a comprehensive cognitive map as described by Tolman (1948).

The guide to the use of technology in the classroom reveals the beneficence of technology in maintenance and facilitation functions aiding the learner in acquiring content knowledge. The induction of new knowledge rests in the sense-full and is thus an individual and social mind phenomenon currently beyond the reach of the *a priori* logic of technology. That said, Kant (1781/1998) does make the case that there is one synthetic *a priori* fundamental to empiricism, every event has a cause. The logic of technology may one day find a way to promote induction of new knowledge. Humans, after all, are the inventors of technology.

References

Alarcón, R. (2009). Culture, cultural factors and psychiatric diagnosis: Review and projections. *World Psychiatry, 8*(3), 131–139. doi:10.1002/j.2051-5545.2009.tb00233.x.

Alexander, P. A. (1996). The past, the present and future of knowledge research: A reexamination of the role of knowledge in learning and instruction. *Educational Psychologist, 31*, 89–92.

Atkin, A. (2013). Peirce's theory of signs. In E. N. Zalta (ed.), *The Stanford encyclopedia of philosophy*. https://plato.stanford.edu/archives/sum2013/entries/peirce-semiotics/.

Atkinson, R., & Shiffrin, R. (1971). The control of short-term memory. *Scientific American, 225*(2), 82–91.

Baddeley, A. (2000). The episodic buffer: A new compartment of working memory? *Trends in Cognitive Sciences, 4*(11), 417–423.

Bateson, G. (1979). *Mind and nature: A necessary unity*. Hampton Press.

Beane, M. (2019). Learning to work with intelligent machines. *Harvard Business Review*, September–October Issue.

Berkeley Center for Teaching & Learning (2020, 14 September). What is learning? https://teaching.berkeley.edu/resources/learn/what-learning.

Brainerd, C. J. (1978). The stage question in cognitive-developmental theory. *The Behavioral and Brain Sciences, 1*, 173–183.

Bransford, J. D., et al., (2006). Learning theories and education: Towards a decade of synergy. In P. A. Alexander & P. Winne (eds.), *Handbook of educational psychology* (2nd ed.), (pp. 209–244). Erlbaum.

Broadbent, J. (2017). Comparing online and blended learner's self-regulated learning strategies and academic performance. *Internet and Higher Education, 33*, 24–32.

Burt, C. (1960), Gustav Theodor Fechner elemente der psychophysik 1860. *British Journal of Statistical Psychology, 13*, 1–10. doi:10.1111/j.2044-8317.1960.tb00033.x.

Campbell, J. (1972). *Myths to live by*. Bantam Books.

Causo, A., Win, P. Z., Guo, P. S., & Chen, I.-M. (2017). Deploying social robots as teaching aid in pre-school K2 classes: A proof-of-concept study. *Proceedings of the 2017 IEEE International Conference on Robotics and Automation (ICRA)*, Singapore, May 29–June 3, 2017.

Dewey, J. (1938). *Experience and education*. Macmillan.

Duval, T. S., & Wicklund, R. A. (1972). *A Theory of objective self-awareness*. Academic.

Emde, R. & Gaensbauer, T. (1981). Some emerging models of emotion in human infancy. In K. Immelmann, G. Barlow, L. Petrinovich, & M. Main (eds.), *Behavioral development: The Bielefeld interdisciplinary project* (pp. 568–588). Cambridge University Press.

Fenigstein, A., Scheier, M., & Buss, A. (1975). Public and private self-consciousness: Assessment and theory. *Journal of Consulting and Clinical Psychology, 43*(4), 522–527.

Fitts, P. (1954). The information capacity of the human motor system in controlling the amplitude of motion. *Journal of Experimental Psychology, 47*(6), 381–391.

Goldberg, E. (2009). *The new executive brain: Frontal lobes in a complex world.* Oxford University Press. ISBN 978–0195329407.
Goldberg, E. (2018). *Creativity: The human brain in the age of innovation.* Oxford University Press.
Goldman-Rakic, P. (1987). Circuitry of primate prefrontal cortex and regulation of behavior by representational memory. In F. Plum & V. Mountcastle (eds.), *Supplement 5 handbook of physiology, the nervous system, higher functions of the brain* (pp. 373–417). American Physiological Society.
Goldman-Rakic, P. (1996). Regional and cellular fractionation of working memory. *Proceedings of the National Academy of Sciences, 93*(24), 13473–13480.
Gottlieb, G. (1998). Normally occurring environmental and behavioral influences on gene activity: From central dogma to probabilistic epigenesis. *Psychological Review, 105*(4), 792–802.
Greene, M. F. (2020, June 23). 30 Years ago, Romania deprived thousands of babies of human contact. *The Atlantic.* https://www.theatlantic.com/magazine/archive/2020/07/can-an-unloved-child-learn-to-love/612253/?utm_source=share&utm_campaign=share.
Greeno, J. G. (2006). Learning in activity. In R. K. Sawyer (ed.), *The Cambridge handbook of the learning sciences* (pp. 79–96). Cambridge University Press.
Haupt, E. (2001). The first memory drum. *The American Journal of Psychology,114*(4), 601–622. doi:10.2307/1423613.
Hofer, M. (1981). *Roots of human behavior.* W. H. Freeman & Company.
Hofer, M. (1984). Relationships as regulators: A psychobiologic perspective on bereavement. *Psychosomatic Medicine, 46*(3), 183–197.
Hofstadter, D. (1979). *Gödel, Escher, Bach: An eternal golden braid* (20th Anniversary Edition). Basic Books.
International Classification of Diseases, 10th Edition (2020). Diagnosis Code F32.9, Major Depressive Disorder, Single Episode, Unspecified.
James, W. (1906–1907).*Pragmatism: A new name for some old ways of thinking.* The Project Gutenberg eBook of Pragmatism, eBook #5116. Released February 2004, updated 2013. https://www.gutenberg.org/files/5116/5116-h/5116-h.htm#link2H_4_0004.
Kant, I. (1998). *Critique of pure reason.* (P. Guyer and A. Wood trans.) Cambridge University Press. (Original work published 1781.)
Kolodner, J. L. (2006). Case-based reasoning. In R. K. Sawyer (ed.), *The Cambridge handbook of the learning sciences* (pp. 225–242). Cambridge University Press.
Kou, Z. (1976). *The dynamics of behavioral development: An Epigenetic view.* Plenum Press.
Loh, E. (2018). Medicine and the rise of the robots: A qualitative review of recent advances in artificial intelligence in health. *British Medical Journal Leader,2,* 59–63. doi: 10.1136/leader-2018-000071.
Lord, A. B. (1971). *The singer of tales.* Atheneum.
Lucas, G., Gratch, J., King, A., & Morency, L.-P. (2014). It's only a computer: Virtual humans increase willingness to disclose. *Computers in Human Behavior, 37,* 94–100.
Lynch, J. (1971). *A single unit analysis of contour enhancement in the somesthetic system of the cat.* Unpublished doctoral dissertation, Stanford University.

Mandler, J. (1979). *Stories, scripts, and scenes: Aspects of schema theory.* Lawrence Erlbaum Associates, Inc.

Mitra, S. & Dangwal, R. (2010). Limits of self-organizing systems of learning—The Kalikuppam experiment. *British Journal of Educational Technology, 41*(5), 672–688.

Ness, J., Kolditz, T., Lewis, P. & Lam, D. (2010). Development and implementation of the U.S. Army leader self-development portfolio. In P. T. Bartone, R. H. Pastel, & M.A. Vaitkus (eds.), *The 71-F advantage: Applying research psychology for health and performance gains* (p. 141). National Defense University Press.

Ness, J. W., Lewis, P., & Brazil, D. (2011). Building the total system: The effects of strong environments on personality development. In *Proceedings of the Human Systems Integration Symposium, "Sharpening the Spear: Integration and Interoperability for Warfighter Effectiveness".* American Society of Naval Engineers.

Norman, D. (2013). *The design of everyday things.* Basic Books.

Oyama, S. (1985). *The ontogeny of information.* Cambridge University Press.

Payer, L. (1996). *Medicine and culture.* Henry Holt & Company.

Paulson, A. (2014, April 21). Blended learning revolution: Tech meets tradition in the classroom. *The Christian Science Monitor Weekly,* retrieved at https://www.csmonitor.com/USA/Education/2014/0420/Blended-learning-revolution-Tech-meets-tradition-in-the-classroom

Peirce, C. S. (1887). The fixation of belief. *Popular Science Monthly, 12*(Nov), 1–15.

Piaget, J. (1978). *Behavior and evolution.* Random House.

Pintrich, P. (2000). The role of goal orientation in self-regulated learning. *Handbook of Self-regulation.* doi:10.1016/B978–012109890-2/50043-3.

Popper, K. R. (1966). *Of clouds and clocks.* Washington University Press.

Pribram, K. (1991). *Brian and perception: Holonomy and structure in figural processing.* Lawrence Erlbaum Associates.

Recanati, F. (2002). Pragmatics and semantics. In L. Horn& G. Ward (eds.), *Handbook of pragmatics.* Wiley-Blackwell.https://jeannicod.ccsd.cnrs.fr/ijn_00000091/document.

Rogoff, B. (1991). *Apprenticeship in thinking.* Oxford University Press.

Scardamalia, M., & Bereiter, C. (2006). Knowledge building: Theory, pedagogy, and technology. In K. Sawyer (ed.), *Cambridge handbook of the learning sciences* (pp. 97–118). Cambridge University Press.

Seigler, R. (1986). *Children's thinking.* Prentice Hall.

Seow, S. (2005). Information theory models of HCI: A comparison of the Hick-Hyman Law and Fitts' Law. *Human-Computer Interaction, 20,* 315–352.

Shannon, C. E. (1948). A mathematical theory of communication. *The Bell System Technical Journal, 27,* 379–423 (July), 623–656 (October).

Silvia, P., & Duval, T. (2001). Objective self-awareness theory: Recent progress and enduring problems. *Personality and Social Psychology Review, 5*(3), 230–241.

Skinner, B. F. (1972). Why I am not a cognitive psychologist. *Behaviorism, 5*(2), 1–10.

Stevens, S. S. (1957). On the psychophysical law. *The Psychological Review, 64*(3), 153–181.

Thayer, J., & Lane, R. (2009). Claude Bernard and the heart–brain connection: Further elaboration of a model of neurovisceral integration. *Neuroscience and Biobehavioral Reviews, 33*(2), 81–88.

The Economist (29 June 2013). E-ducation. *The Economist Newspaper Limited*, retrieved at https://www.economist.com/leaders/2013/06/29/e-ducation.

Thurstone, L. (1928). Attitudes can be measured. *American Journal of Sociology, 33*, 529–554.

Tolman, C. (1948). Cognitive maps in rats and men. *Psychological Review, 55*(4), 189–208.

Turgue, B. (2012, March 6). 'Creative....Motivating' and fired. *Washington Post*. https://www.washingtonpost.com/local/education/creative--motivating-and-fired/2012/02/04/gIQAwzZpvR_story.html.

U.S. Department of Education—Office of Planning, Evaluation, and Policy Development (2009). *Evaluation of evidence-based practices in online learning: A meta-analysis and review of online learning studies.*

Vygotsky, L. (1988). *Thought and language* (3rd printing). The MIT Press.

Waddington, C. H. (1966). *Principles of development and differentiation.* Macmillan Company.

Wasdyke, C. & Ness, J. W. (2019). *Lexicon of mood: Ethical implications for intelligent agent design.* DoD Human Factors Engineering Technical Advisory Group Meeting 73, Aberdeen Proving Ground, MD.

Whitehead, A. N. (November 1919). The concept of nature. In J. Kegg & L. Wisewell (eds.), *Project Gutenberg*, released 16 July 2006, eBook #18835. https://www.gutenberg.org/files/18835/18835-h/18835-h.htm.

Wicklund, R. & Gollwitzer, P. (1987). The fallacy of the private-public self-focus distinction. *Journal of Personality, 55*(3), 491–523.

Wittgenstein, L. (2010). *Tractatus logico-philosophicus eBook.* Project Gutenberg (Original work published 1921).https://www.gutenberg.org/files/5740/5740-pdf.pdf.

Zal'mason. (1926). *Concerning the condition of excitation in dominance.* Nova y refteksolgie I fizologi.

Chapter 5

Benefits and Potential Issues for Intelligent Tutoring Systems and Pedagogical Agents

Lishan Zhang, Xiangen Hu, Frank Andrasik, and Shuo Feng

1 Introduction

Intelligent agents are becoming an essential part of our everyday lives. They continuously sense a human's requirements and automatically select and execute appropriate actions to accomplish specific tasks (Franklin & Graesser, 1996). They can help people manage their schedules and contacts, control the smart equipment in the room, order meals from restaurants, and other tasks that previously have been solely the domain of human agents. On one hand, intelligent agents facilitate many routine tasks and make our lives increasingly more convenient. On the other hand, development of agents may raise serious ethical issues. At present, the effects that agents will have on people's thoughts and beliefs, particularly over the long-term, are unclear.

Of particular concern is that the intelligent agents introduced above were not originally developed to train or educate people. Rather, they were designed to help humans finish some specialized tasks. The effects on human's beliefs are perhaps best thought of as unintended side effects as we pursue convenience and efficiency. This fact compels us to be concerned about the uncertain effects of *pedagogical* intelligent agents that are usually situated in intelligent tutoring systems (ITSs), and designed to educate people through a series of interactions. ITSs are designed to provide learners with personalized learning experiences to improve their learning efficiency. Sometimes, the intelligent tutoring system itself is a pedagogical agent; for example, in conversation-based intelligent tutoring systems. Evaluations typically consider human experts as the gold standard, and ITSs often mimic how human experts teach learners. Consequently, ITS applications are judged as successful if learners cannot tell the difference between ITSs and human experts. However, in this chapter, we take the position that it is not always appropriate or necessary to disguise ITSs as human agents and make learners believe they are interacting with real human tutors instead of a software agent.

DOI: 10.4324/9781003030928-8

We begin by briefly expanding upon the functions of a pedagogical agent, which is the intelligent agent responsible for conducting the communication between learners and conversational ITSs. Then we summarize the roles that ITSs can serve to learners to help them see the various ways that ITSs can be used productively. We next turn our attention to the topic of technology acceptance in education in order to understand the concerns that different user populations may have toward ITSs. We conclude with a discussion, wherein we outline aspects that we believe warrant serious consideration when designing an ITS.

2 Pedagogical Agents in ITSs

An ITS is a system that can help learners overcome misconceptions, scaffold learning, and develop needed competencies by implementing customized instruction. A typical ITS consists of four components: a domain model, learner model, pedagogical model, and user interface. Learners interact with an ITS through a user interface, acquire the knowledge provided by the domain model, and are tutored by the pedagogical model which implements the desired instructional methods. While learners are interacting with the ITS, the entire learning process is tracked and analyzed, so that the learner model is updated and reflects the current learning state.

In the domain of intelligent tutoring systems, the pedagogical agent refers to a human-like character simulated by the ITS (Chou, Chan, & Lin, 2003). From the perspective of learners, the pedagogical agent is part of the user interface. Learners can interact with the ITS by engaging with the agent in a simulated and virtual representation of some roles, such as teachers, coaches, mentors, or even peer learners.

A pedagogical agent presents as a virtual character with some human-like traits and functions, primarily by simulating human appearance and/or behavior (Levillain & Zibetti, 2017). Graphic capabilities and simulated technologies have developed such that they are no longer limited to two-dimensional space or text-only (Damer, 1997; Roschelle, Feng, Murphy, & Mason, 2016; Williams et al., 2016). In fact, many ITS user interfaces now incorporate three-dimensional environments and agents to achieve a highly advanced simulated effect (Bishop, 2010; Lane, Noren, Auerbach, Birch, & Swartout, 2011; Liao, Sunq, Wang, & Lin, 2019; Talamo & Ligorio, 2001).

To make pedagogical agents better equipped to understand humans, in addition to cognitive states, affective computing can detect the affective states of learners from their behaviors, which enables the pedagogical agents to provide immediate personalized feedback (D'Mello et al., 2008). As a further step, some systems, for example, Affective

AutoTutor (D'Mello et al., 2008), have enabled the pedagogical agent to simulate human facial expressions so that interactions with the learners include emotional elements, which is a highly beneficial and profound breakthrough in both appearance and behaviors. In short, the essence of the development of a pedagogical agent is to endow non-human agents with "human" characteristics, or anthropomorphism. The remainder of this section is devoted to discussion of anthropomorphism in greater detail and the effects of a pedagogical agent's anthropomorphism.

2.1 Anthropomorphism

All intelligent agents inherently utilize anthropomorphism. Conventionally, over-application of anthropomorphism was regarded as a conception error common to immature or naive people, notably children (Caporael, 1986; Fisher, 1996; Mitchell, 2005). Sherry Turk, for example, conceptualizes anthropomorphism as creating an illusionary relationship, one that potentially engenders misunderstandings that carry over into human relationships. Such a perspective condemns the application of anthropomorphic technology within the context of a social activity, on the grounds of adulterating interpersonal communication (Allen & Wallach, 2012).

However, other researchers maintain that anthropomorphism should not be constrained because its application to ITSs constitutes simply an in-kind expansion of standard practices that predate modern technology. Anthropomorphism, they argue, makes articulations vivid, impressive, and funny (Damiano & Dumouchel, 2018). Consider the example of imbuing flowers with human activity in the description of them "dancing with the wind". Anthropomorphism, then, is a basic function of the mind, not an error of cognition (Złotowski, Proudfoot, Yogeeswaran, & Bartneck, 2015). Further, with the development and advancement of artificial intelligence, the distinction between human and robot blurs. It may soon prove untenable to maintain strict dualism that stresses the need to distinguish human from non-human, organic from artificial. A synthetic perspective may prove inescapable if we are to successfully consider the relationship between human and intelligent anthropomorphic entities. As artificial intelligence increasingly and dramatically impacts human life, the process of *coevolution* may prove the more reasonable course (Damiano & Dumouchel, 2018).

In summary, researchers hold different perspectives regarding anthropomorphic features of intelligent agents. Some believe that intelligent agents should be clearly identified especially when they are used in a social activity, but others believe that a human's existing tendency to anthropomorphize should be extended and leveraged to foster effective collaboration with machines in the emerging society.

2.2 Effects of Pedagogical Agents

Whether a pedagogical agent is clearly identified or not, it impacts social functions while interacting with learners. Thus, it is important and necessary to study its effects on learners. The remainder of this section discusses the effects from three perspectives: presence and motivation, cognitive load, and social interactions.

2.2.1 Presence and Motivation

Presence is the sense of being there, a central feature of a pedagogical agent. More specifically, presence contains telepresence (users feel immersed in a virtual environment), co-presence (users feel accompanied by others) and social presence (users feel accessed to others' minds). Nowak and Biocca (2003) found that the virtual character can increase telepresence, but evidence showing that a virtual pedagogical agent is beneficial for providing co-presence and social presence is limited. In fact, a pedagogical agent with a high degree of anthropomorphism may hinder learners' co-presence and social presence (Nowak & Biocca, 2003).

When learners are able to feel the existence of the pedagogical agent, the agent has the opportunity to motivate the learners. For example, the pedagogical agent can help learners to set learning goals, make plans for achieving those goals together, and periodically remind the learners to follow their plans (Aleven, Roll, McLaren, & Koedinger, 2016). In this way, the learners are motivated and self-regulated, so learning efficiency can be improved.

2.2.2 Cognitive Load

It is also important to study learners' cognition costs when interacting with a pedagogical agent, to better understand demand distribution during the interaction. The theory of cognitive load argues that working memory (short-term memory) load is the combination of intrinsic (the work itself), extraneous (unrelated to work and schema), and germane (related to the schema) cognitive load. From this conception, germane cognitive load should be increased and extraneous load decreased when designing learning systems to optimize learning relative to effort (Kester, Lehnen, Van Gerven, & Kirschner, 2006).

This has bearing on the design of anthropomorphic qualities of a pedagogical agent. Approximating human characteristics requires some features unrelated to learning (Clark & Choi, 2005; Woo, 2009). For example, the tendency for a pedagogical agent to blink lacks intrinsic learning value. However, such features may distract learners' attention, potentially increase extraneous cognitive load, and disrupt the learning performance. Consequently, pedagogical agents sometimes need to sacrifice anthropomorphism in order to enable learners to concentrate on learning, as opposed to simply enjoying "playing" with the pedagogical agent.

2.2.3 Social Interaction

From a sociological perspective, learning is a social activity occurring in a particular context. Here, the pedagogical agent participates in and can influence learners in attaining knowledge, improving skills, feeling emotions, and even forming ethics (Veletsianos & Russell, 2014). When the pedagogical agent is also self-improvable, it can form human–agent coevolution, with both parties adapting independently and in response to the other. Establishing this dynamic likely constitutes a major development in the gradual blurring of borders between humans and agents introduced above. Thus, it is important to understand the mechanisms of how social activities are influenced by the combination of humans and agents, and to distinguish their effects.

3 The Roles that ITSs Can Play Social Interaction

The design and implementation of a practical ITS is usually domain dependent, which means that each ITS typically specializes in only one teaching domain. For example, *ASSISTments*, *Cognitive Tutor*, and *Thinkster Math* teach mathematics (Maths, 2020; Ritter, Anderson, Koedinger, & Corbett, 2007; Roschelle et al., 2016), *Betty's Brain* teaches science (Biswas, Segedy, & Bunchongchit, 2015), *Dragoon* teaches dynamic modeling (Vanlehn, Wetzel, Grover, & Sande, 2017) AEINS teaches ethics (Hodhod, Kudenko, & Cairns, 2009), and *Duolingo* teaches language (Von Ahn, 2013).

To motivate learners and improve their learning efficiencies in the designed teaching domain, ITSs do not always behave only as software tools. Some ITSs implement virtual agents that play the roles of teachers, peers that need to be tutored, and classmates. Based on the designated roles of the virtual agents, learners can build different relationships with the ITS. The remainder of this section reviews the typical roles that an ITS and its virtual agents can play to learners.

3.1 ITSs as Software Tools

Some ITSs do not use anthropomorphism. Although they may use some artificial intelligence or machine learning technologies, they purely behave as software tools. For example, *ASSISTments* is a tool to help learners finish their mathematics homework online. It can give immediate feedback to learners (Roschelle et al., 2016), helping prevent them from making the same type of mistakes repeatedly. At the same time, the system generates assessment reports of all the learners as well as each individual learner for teachers, so that the teachers can revise their future instructions accordingly. In other words, the ITS combines the assistance with assessment—hence the name. A series of studies on ASSISTments have shown that it can increase learners' academic achievement and enable formative assessment for teachers.

3.2 ITSs that Implement Teacher Behaviors

Some ITSs not only provide intelligent instructions, but they also look and behave like real teachers. In this case, learners typically use natural language to interact with the virtual teacher and learn the related concepts by conducting dialogs. This basic approach can have disparate instantiations. Here we review *AXIS* and *AutoTutor* as illustrations.

In a traditional classroom, teachers can overview and summarize learners' explanations for a problem, and may select some as references for class. In this approach, both learners and teachers would identify misunderstandings and know what to do next. AXIS (Adaptive eXplanation Improvement System) adopts this teacher role to provide suitable explanations for learners. It adopts a crowdsourcing method to let learners generate, revise, and evaluate explanations by themselves for future learners on the platform (Williams et al., 2016). Unlike the usual methods of teachers or tutors inputting explanations, crowdsourcing may scaffold and leverage them considerably, because learners are experts in their typical misconceptions. Besides, when learners generate, revise, and evaluate explanations, they will obtain a deeper sense of the problem.

In the domain of cognition and metacognition, their abilities to understand, summarize, read and self-reflect will also improve. These are crucial for future learning. The system has two core components: the student sourcing interface and the explanation selection policy. The learners can write, receive, and evaluate explanations with scale through the student sourcing interface. The explanation selection is the policy of deciding which explanation will be shown to future learners. The system adopts Thompson sampling, a reinforcement learning algorithm, to address the problem of exploitation versus exploration. Follow-up studies have shown that the learning gains and quality of explanations of AXIS do not differ from those written by teachers.

AutoTutor instructs by conducting a series of dialogs. The original AutoTutor has developed into a family of systems covering many domains, including physics, computer science, and scientific method (D'Mello & Graesser, 2012). AutoTutor simulates a human tutor with a pedagogical agent, communicating with learners in natural language and using their input to monitor cognitive states. Ideally, when learners interact with AutoTutor, they feel as if they are talking with a virtual tutor instead of using a computer program. To enable the virtual tutor to behave much like a real human tutor, researchers constructed dialog structure templates through observation of live actual classroom instruction. This structure follows a five-step tutoring frame. The five steps are (a) the agent formulates a question for the learner to discuss; (b) the learner thinks about the question and answers the question, thereby indicating their level of conceptual mastery; (c) the agent receives the answer, identifying components of a correct answer as well as potential misconceptions; (d) the agent provides hints or prompts to expand and refine the answer; and (e) finally, the agent tests whether the learner really understands the original question (D'Mello & Graesser, 2012). This process is termed the learning-by-talking method, where the agent and the learner collaboratively construct the comprehensive answer together, and learners acquire and strengthen the critical concepts.

3.3 ITSs that Implement Classmate Behaviors

In a traditional class, classmates play an important role because learners can communicate and solve problems collaboratively, contributing to successful learning (Walton, Cohen, Cwir, & Spencer, 2012). Conversely, learners can easily feel isolated when using online education platforms (Dalipi, Imran, & Kastrati, 2018). To reduce learners' loneliness, some virtual agents are designed to mimic classmates' behaviors, especially the behaviors in chatrooms and forums (Coetzee, Lim, Fox, Hartmann, & Hearst, 2015; Kulkarni, Cambre, Kotturi, Bernstein, & Klemmer, 2015). *Virtual Classmates* (Liao et al., 2019) is an example of an ITS using virtual reality technologies for online education.

In Virtual Classmates, human learners watch online course videos along with virtual classmates empowered by machine learning technologies. To make virtual classmates behave as human learners, each virtual classmate is assigned a unique personality. The virtual classmates then automatically comment on the course videos based on the assigned personality. The existing comments on course videos from human learners are used as the training data, so that the machine intelligence can be implemented. A Naïve Bayes classifier is trained to map the comments to virtual classmates and make the comments appear at reasonable time points in the course videos. As a result, while learners are watching the course videos, they will see some comments like "I wanted to learn this lesson before" and "I like the music at the beginning of the lesson". This creates the illusion that peers are watching the course videos alongside the learner. Thus, the virtual agents can potentially reduce the learners' perceived isolation and help keep them motivated.

3.4 ITSs that Implement Learner Behaviors

The role of the virtual agent can also extend further, serving as a peer learner whose understanding of the content is dynamic. In this case, the theory of learning-by-teaching is implemented to enable learners to teach virtual agents and therefore strengthen their own knowledge through conducting the instruction. In other words, learners become teachers in this paradigm, and they do not need to take assessments anymore. However, they do need to correctly teach the virtual agents, so that the agents can pass the assessments. This paradigm makes learners feel responsible for the virtual teachable agents. When learners have trouble correctly teaching the virtual agents, they actively learn by themselves and explain to the virtual agents what they have learned. The learning and explaining process both help learners construct deep understanding on the related concepts. Below, the ITSs *Betty's Brain* and *SimStudent* illustrate how the theory of learning-by-teaching is implemented.

Betty's Brain provides an open-ended learning environment where human learners can acquire science knowledge, through reading texts and teaching the virtual agent, Betty (Biswas et al., 2015). More specifically, learners need to construct the concept map of Betty, so that Betty can correctly answer the assessment questions. A concept map is a form of knowledge representation which describes the relation of diverse concepts in complex tasks and contexts. The concept map actually represents Betty's as well as the learners' domain knowledge. The correctness of Betty's assessments also reflects the learners' understanding. The learners can then learn by revising the concept map according to the feedback on Betty's assessments. In addition to Betty, another virtual agent, named Mr. David, is implemented to provide metacognitive suggestions during the human learners' teaching. Mr. David can remind learners to carefully revise Betty's concept map and send Betty to take the assessment. Human learners can learn more effectively when they are regulated to follow these teaching strategies.

SimStudent is also a teachable agent, often used in algebra learning and tutoring. It applies machine learning engines to learn principles from the examples that users provide (Matsuda, Cohen, & Koedinger, 2005). SimStudent can serve two roles: as an authoring tool and a learning-by-teaching paradigm (Matsuda, Cohen, & Koedinger, 2014; Matsuda, Cohen, Sewall, Lacerda, & Koedinger, 2007; Matsuda et al., 2013). As an authoring tool, it can help novice authors develop cognitive tutors easily (Matsuda et al., 2014). Authors can interact with it to author ITS expert models through tutoring or demonstration without complex programming. As a virtual student, it uses a learning-by-teaching method to enhance learning in both cognitive and metacognitive levels. In this role it focuses on procedural algebra problem solving and is integrated in APLUS (Artificial Peer Learning environment Using SimStudent), a feature-rich and game-like online system (Matsuda et al., 2013). It employs demonstrations to generalize examples and form rules. More specifically, learners need to create a problem and allow the agent to answer it first. Based on the answers, learners need to provide feedback to tell the agent right or wrong. When the agent is stuck, learners can provide help and explanations. In addition, learners can have the agent take quizzes to see the learning effect. During the tutoring process, the agent can interact with learners in the form of a virtual avatar, which allows the agent to show some facial expressions and emotional states.

4 Digital Technology Acceptance in Education

Although intelligent tutoring systems are arguably becoming more diverse and effective, the end users may weigh the importance of those practical values differently when they are enabled by unfamiliar advanced technologies. Over the years, some well-known models have been built to explore people's digital technology acceptance. The Technology Acceptance Model (TAM) (Davis, Bagozzi, & Warshaw, 1989) and the Unified Theory of Acceptance and Use of Technology (UTAUT) (Venkatesh, Morris, Davis, & Davis, 2003) are two that have been widely used. This section summarizes some important studies on digital technology acceptance in education with the help of TAM and UTAUT from three different perspectives: teacher, learner, and parent. Reviewing these works allows us to gain insights into and predict issues likely to arise when intelligent tutoring systems are introduced in everyday teaching practice.

4.1 Technology Acceptance Model Roles

TAM is commonly used for a broad range of target groups and technologies (e.g., Hsiao & Yang, 2011; Ifenthaler & Schweinbenz, 2016). It originally defined two main constructs: perceived usefulness (the degree of feeling helpful for users when adopting technology to work) and perceived ease of use (the degree of feeling effortless for users when adopting technology to work). Furthermore, Davis (1989) took other variables (i.e., behavioral intention to use, actual system use, and attitudes toward technology) along with these two mentioned here to establish a causal system. The causal relation of these variables is illustrated in Figure 5.1.

UTAUT is another powerful model that merits mention. This model defines four core influence factors: effort expectancy (the degree of feeling effortless when using technology), performance expectancy (the degree of feeling helpful for work), social influence (the degree of perceiving that others tend to use technology), and facilitating conditions (the degree of feeling support from technological infrastructure), behavioral

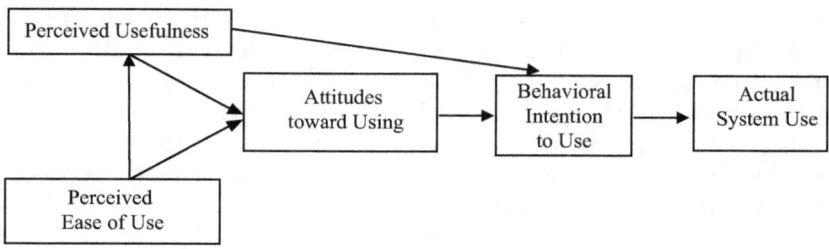

Figure 5.1 Technology Acceptance Model (TAM).

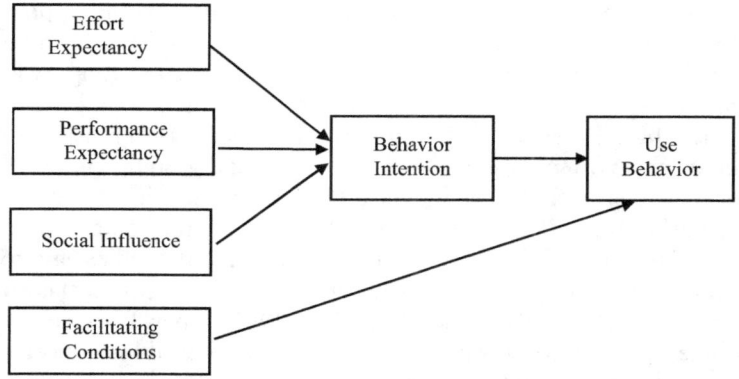

Figure 5.2 Unified Theory of Acceptance and Use of Technology (UTAUT).

intention, and use behavior (Venkatesh et al., 2003). The casual relationship of this model is illustrated in Figure 5.2. As these figures illustrate, the conceptualization of determinants in this model is similar to TAM (Nistor & Heymann, 2010). These determinants can also be moderated by user characters, such as gender, age, experience, and attitude (Williams, Rana, & Dwivedi, 2015), which makes the model more difficult to be evaluated than TAM (Scherer, Siddiq, & Tondeur, 2019).

4.2 Teacher's Digital Technology Acceptance

Teachers play an important role in integrating technology into instruction, and are the ones who conduct the actual instruction. The degree of a teacher's acceptance can vary based on personal characteristics. Many studies have been conducted to disclose the factors that affect a teacher's acceptance, with the three below surfacing as particularly important.

The first is gender. Researchers introduced gender as a potential moderating factor into UTAUT, and found it had a significant moderating effect in an empirical study (Im, Kim, & Han, 2008). Im et al. showed that a male teacher's acceptance was more affected by perceived usefulness and less affected by perceived ease of use. This implies that teachers who are males can be attracted more by the complex technology that provides broad functionality, while teachers who are female favor technology with easy-to-learn user interface.

Another important factor is age, which is typically a reflection of teaching experience. Senior teachers often have already formed and are comfortable with their pedagogical habits and, thus, are less motivated to pursue opportunities for new learning and advancement. On the other hand, young teachers are often more interested in and open to trying out new and different possibilities in teaching, and also seeking opportunities for advancement. Teachers who are younger have been found to be more willing to accept using new technology in their teaching. Another explanation for age serving as an important factor is its high correlation with digital literacy (O'Bannon & Thomas, 2014; Prensky, 2001). Young teachers often have less difficulty in using digital technology than senior teachers, so they have higher digital acceptance in education.

Third, the conception of teaching and learning can also affect a teacher's digital technology acceptance (Teo & Zhou, 2016). Studies have found that constructivist teachers tend to incorporate technology into classrooms more often (Becker, 1999; Becker & Riel, 2000; Overbay, Patterson, Vasu, & Grable, 2010). Teo and Zhou (2016) revised TAM and introduced more factors including subjective norm, facilitating condition, conception of learning and teaching, and traditional and constructivist conception of learning and teaching. In doing so, they found that conception of learning and teaching had a significant influence on technology acceptance, with the constructivist teachers showing higher acceptance of technology.

4.3 Learner's Digital Technology Acceptance

Although learners are usually open to accepting digital technologies due in large part to curiosity, their acceptance is not unified and is affected by attitude and behavioral factors. Ifenthaler and Schweinbenz (2016) studied learners' acceptance of tablet personal computers in three German middle schools. Their research model, which integrates TAM and UTAUT, led them to hypothesize that:

a Social influence (SI) and self-efficacy (SE) influence behavioral intentions (BI) directly.
b Effort expectancy (EE), performance expectancy (PE), and social influence (SI) affect attitudes toward using (ATU).
c Facilitating conditions (FC) have an effect on use of technology (UT).

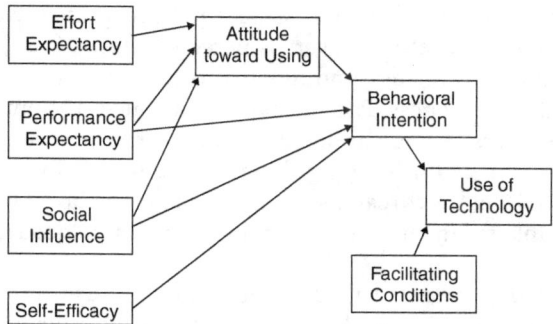

Figure 5.3 Research model for learner digital acceptance.

The model is illustrated in Figure 5.3. Their study did find that effort expectancy and performance expectancy affect attitude positively, but that social influence did not significantly affect attitude toward using and behavioral intentions. Further, self-efficacy influence did not significantly influence behavioral intention, and facilitating conditions had no relation to use of technology. In summary, the results showed that learners tended to accept less-effort technology and that feeling helpful when learning was crucial for them. Other non-significant effects merit further exploration due to their novelty and other limitations.

As was true for teachers (in earlier studies mentioned), male and female learners showed different degrees of digital technology acceptance. Nami and Vaezi (2018) confirmed the argument based on a study conducted in a university. Their research model added perceived self-efficacy as a third seminal construct to extend the two embodied in TAM, perceived ease of use and usefulness. Their study additionally investigated learners' technology knowledge in another scale. They found that male learners had higher self-efficacy about using digital technology. Further, learners who possessed a personal computer reported feeling more comfortable using technology than those who do not have or access to a personal computer.

4.4 Parent's Digital Technology Acceptance

Parents have a profound influence on learners, whose usage of digital technology often needs to be under the permission of their parents. Zhu et al. (2014) explored the acceptance of tablet usage in class in a study comparing parents in the US and China. Building upon the two key constructs in TAM, perceived ease of use and perceived usefulness, they introduced other factors they believed would be helpful in explaining acceptance from the perspective of parents.

They found that the challenge of parents' acceptance was governed mainly by perceived usefulness versus perceived ease of use. Parents believed that tablets were easy and flexible to use, but they doubted their efficiency and effectiveness. Parents expressed concern about the Internet distracting learners. Some parents even stated they perceived no benefit at all to using tablets for learning. Finally, parents also voiced concerns about potential health threats, such as myopia. In short, parents often appeared unable to understand why tablet computers needed to be used in class.

Previous studies revealed that efficient communication can mitigate parents' worries. Tsuei and Hsu (2019) introduced four variables (i.e., self-efficacy, parents' belief, parent–teacher communication, and parent–children interaction) as external variables in TAM to address this aspect. They found that parents' belief did affect learner usage of technology at home, and that both parent–teacher communication and parent–child communication enhanced the understanding of parents with respect to technology application in education.

5 Discussion

In this chapter, we reviewed the potential effects of pedagogical agents in intelligent tutoring systems and the different roles that agents can play. We then briefly summarized people's digital technology acceptance in education. Having done this allows us to infer general concerns that pertain to intelligent tutoring systems.

From the existing studies, it is clear that pedagogical agents can not only directly affect cognition with respect to learning content, but they also can affect learners' social interactions. Assuming this effect spreads via social interactions, it becomes increasingly intractable to distinguish changes in social cognitive patterns when interacting with pedagogical agents, other humans, or both. This situation has potential for creating serious ethical problems. For example, if a learner incorporates ideas or conclusions outside of the expected and approved curriculum, who is the responsible party? In reality, it will likely not even be possible to identify the source of the issue.

In addition, state of the art pedagogical agents can behave as roles other than a tutor. We can envision a future where learning occurs in a setting that combines human and artificial intelligence classmates, such as a lecture delivered by a group of human and artificial intelligence teachers. The problem to confront then will be whether a human learner can always distinguish intelligent agents from the combinations and what if some human learners are in favor of learning with intelligent agents rather than real humans. Might the differing preferences affect human learners' social interactions? It seems important to tag these intelligent agents, so that learners are always clear about from whom they are learning.

People in general, but especially the parents of learners, are rightfully concerned about how to work with digital technologies in education and their unexpected effects, largely due to a lack of understanding. Given that the mechanisms of pedagogical agents are far more complicated than commonly available digital technology, people may be more reluctant to accept them without appropriate context. Therefore, it is the responsibility of researchers to make the underlying mechanisms as transparent as possible, so that teachers, learners, and parents can understand the behaviors of pedagogical agents just as they do for the behaviors of real humans. We hereby offer four suggestions about the ethically responsible usage of pedagogical agents:

1 Pedagogical agents should be clearly identified, instead of being "disguising" as human pedagogical agents.
2 The effect of pedagogical agents should be tractable, so that the agents can be quantitatively evaluated.
3 The mechanisms of pedagogical agents' behaviors should be white box instead of black box, so that people can understand why the agents conduct each behavior.
4 The advantages and disadvantages of each type of pedagogical agent should be carefully evaluated and recorded to make clear the suitability of the context for applying the agents.

We hope that these suggestions will help in building people's trust toward intelligent tutoring systems and the pedagogical agents used. We hope, as well, that a framework can be gradually constructed to explain when and where intelligent tutoring systems are most suitable for application.

Acknowledgements

The work was supported in part by the National Natural Science Foundation of China [Grant Number 61807004]

References

Aleven, V., Roll, I., McLaren, B. M., & Koedinger, K. R. (2016). Help helps, but only so much: Research on help seeking with intelligent tutoring systems. *International Journal of Artificial Intelligence in Education*, 26(1), 205–223.

Allen, C., & Wallach, W. (2012). Moral machines: Contradiction in terms or abdication of human responsibility? In P. Lin, K. Abney, & G. A. Bekey (eds.), *Robot ethics: The ethical and social implications of robotics* (pp. 55–68). MIT Press.

Becker, H. J. (1999). Internet use by teachers: Conditions of professional use and teacher-directed student use. *Teaching, Learning, and Computing: 1998 National Survey. Report# 1*.

Becker, H. J., & Riel, M. M. (2000). Teacher professional engagement and constructivist-compatible computer use. *Teaching, Learning, and Computing: 1998 National Survey. Report# 7*.

Bishop, J. (2010). The role of multi-agent social networking systems in ubiquitous education: Enhancing peer-supported reflective learning. In T. T. Goh (ed.), *Multiplatform e-learning systems and technologies: Mobile devices for ubiquitous ICT-based education* (pp. 72–88): IGI Global.

Biswas, G., Segedy, J. R., & Bunchongchit, K. (2015). From design to implementation to practice a learning by teaching system: Betty's Brain. *International Journal of Artificial Intelligence in Education, 26*(1), 350–364. doi:10.1007/s40593-015-0057-9.

Caporael, L. R. (1986). Anthropomorphism and mechanomorphism: Two faces of the human machine. *Computers in Human Behaviour, 2*(3), 215–234. doi:10.1016/0747-5632(86)90004-x.

Chou, C.-Y., Chan, T.-W., & Lin, C.-J. (2003). Redefining the learning companion: The Past, present, and future of educational agents. *Computers & Education, 40*(3), 255–269.

Clark, R. E., & Choi, S. (2005). Five design principles for experiments on the effects of animated pedagogical agents. *Journal of Educational Computing Research, 32*(3), 209–225.

Coetzee, D., Lim, S., Fox, A., Hartmann, B., & Hearst, M. A. (2015). *Structuring interactions for large-scale synchronous peer learning*. Paper presented at the Proceedings of the 18th ACM Conference on Computer Supported Cooperative Work & Social Computing.

Dalipi, F., Imran, A. S., & Kastrati, Z. (2018). *MOOC dropout prediction using machine learning techniques: Review and research challenges*. Paper presented at the 2018 IEEE Global Engineering Education Conference (EDUCON).

Damer, B. (1997). *Avatars!; Exploring and building virtual worlds on the internet*. Peachpit Press.

Damiano, L., & Dumouchel, P. (2018). Anthropomorphism in human-robot co-evolution. *Frontiers in Psychology, 9*, 468. doi:10.3389/fpsyg.2018.00468.

Davis, F. D. (1989). Perceived usefulness, perceived ease of use, and user acceptance of information technology. *MIS quarterly, 13*(3), 319–340.

Davis, F. D., Bagozzi, R. P., & Warshaw, P. R. (1989). User acceptance of computer technology: A comparison of two theoretical models. *Management Science, 35*(8), 982–1003.

D'Mello, S., & Graesser, A. C. (2012). AutoTutor and affective AutoTutor. *ACM Transactions on Interactive Intelligent Systems, 2*(4), 1–39. doi:10.1145/2395123.2395128.

D'Mello, S., Jackson, T., Craig, S., Morgan, B., Chipman, P., White, H., Person, N., Kort, B., El Kaliouby, R., Picard, R., & Graesser, A. (2008). AutoTutor detects and responds to learners' affective and cognitive states. Paper presented at the *Workshop on Emotional and Cognitive Issues at the International Conference on Intelligent Tutoring Systems*, 306–308.

Fisher, J. A. (1996). The Myth of Anthropomorphism. In M. Bekoff & D. Jamieson (eds.), *Readings in animal cognition* (pp. 3–16). Cambridge: MIT.

Franklin, S., & Graesser, A. (1996, August). Is it an Agent, or just a Program?: A Taxonomy for Autonomous Agents. In *International Workshop on Agent Theories, Architectures, and Languages* (pp. 21–35). Berlin, Heidelberg: Springer.

Hodhod, R., Kudenko, D., & Cairns, P. (2009). *AEINS: Adaptive educational interactive narrative system to teach ethics.* Paper presented at the AIED 2009: 14th International Conference on Artificial Intelligence in Education Workshops Proceedings.

Hsiao, C. H., & Yang, C. (2011). The intellectual development of the technology acceptance model: A co-citation analysis. *International Journal of Information Management, 31*(2), 128–136.

Ifenthaler, D., & Schweinbenz, V. (2016). Students' acceptance of tablet PCs in the classroom. *Journal of Research on Technology in Education, 48*(4), 306–321. doi:10.1080/15391523.2016.1215172.

Im, I., Kim, Y., & Han, H.-J. (2008). The effects of perceived risk and technology type on users' acceptance of technologies. *Information & Management, 45*(1), 1–9.

Kester, L., Lehnen, C., Van Gerven, P. W., & Kirschner, P. A. (2006). Just-in-time, schematic supportive information presentation during cognitive skill acquisition. *Computers in Human Behavior, 22*(1), 93–112.

Kulkarni, C., Cambre, J., Kotturi, Y., Bernstein, M. S., & Klemmer, S. R. (2015). *Talkabout: Making distance matter with small groups in massive classes.* Paper presented at the Proceedings of the 18th ACM Conference on Computer Supported Cooperative Work & Social Computing.

Lane, H. C., Noren, D., Auerbach, D., Birch, M., & Swartout, W. (2011). *Intelligent tutoring goes to the museum in the big city: A pedagogical agent for informal science education.* Paper presented at the International Conference on Artificial Intelligence in Education.

Levillain, F., & Zibetti, E. (2017). Behavioral objects: The rise of the evocative machines. *Journal of Human-Robot Interaction, 6*(1), 4–24.

Liao, M.-Y., Sunq, C.-Y., Wang, H.-C., & Lin, W.-C. (2019). *Virtual classmates: Embodying historical learners' messages as learning companions in a VR classroom through comment mapping.* Paper presented at the 2019 IEEE Conference on Virtual Reality and 3D User Interfaces.

Maths, T. (2020). *Our elite maths coaches can visualize student thinking.* Retrieved from https://hellothinkster.co.in.

Matsuda, N., Cohen, W. W., & Koedinger, K. R. (2005). Applying programming by demonstration in an intelligent authoring tool for cognitive tutors. *Human-Computer Interaction Institute*, 245.

Matsuda, N., Cohen, W. W., & Koedinger, K. R. (2014). Teaching the teacher: Tutoring SimStudent leads to more effective cognitive tutor authoring. *International Journal of Artificial Intelligence in Education, 25*(1), 1–34. doi:10.1007/s40593-014-0020-1.

Matsuda, N., Cohen, W. W., Sewall, J., Lacerda, G., & Koedinger, K. R. (2007). Predicting students' performance with simstudent: Learning cognitive skills from observation. *Frontiers in Artificial Intelligence and Applications, 158,* 467.

Matsuda, N., Yarzebinski, E., Keiser, V., Raizada, R., Cohen, W. W., Stylianides, G. J., & Koedinger, K. R. (2013). Cognitive anatomy of tutor learning: Lessons learned with SimStudent. *Journal of Educational Psychology, 105*(4), 1152.

Mitchell, S. D. (2005). Anthropomorphism and cross-species modeling. In L. Daston & G. Mitman (eds.), *Thinking with animals: New perspectives on anthropomorphism* (pp. 100–118). Columbia University Press.

Nami, F., & Vaezi, S. (2018). How ready are our students for technology-enhanced learning? Students at a university of technology respond. *Journal of Computing in Higher Education, 30*(3), 510–529. doi:10.1007/s12528-018-9181-5.

Nistor, N., & Heymann, J. O. (2010). Reconsidering the role of attitude in the TAM: An answer to Teo (2009a). *British Journal of Educational Technology, 41*(6), E142–E145.

Nowak, K. L., & Biocca, F. (2003). The Effect of the agency and anthropomorphism on users' sense of telepresence, copresence, and social presence in virtual environments. *Presence: Teleoperators & Virtual Environments, 12*(5), 481–494.

O'Bannon, B. W., & Thomas, K. (2014). Teacher perceptions of using mobile phones in the classroom: Age matters! *Computers & Education, 74,* 15–25.

Overbay, A., Patterson, A. S., Vasu, E. S., & Grable, L. L. (2010). Constructivism and technology use: Findings from the IMPACTing leadership project. *Educational Media International, 47*(2), 103–120.

Prensky, M. (2001). Digital natives, digital immigrants. *On the Horizon, 9*(5), 1–6.

Ritter, S., Anderson, J. R., Koedinger, K. R., & Corbett, A. (2007). Cognitive tutor: Applied research in mathematics education. *Psychonomic Bulletin & Review, 14*(2), 249–255. doi:10.3758/bf03194060.

Roschelle, J., Feng, M., Murphy, R. F., & Mason, C. A. (2016). Online mathematics homework increases student achievement. *AERA Open, 2*(4). doi:10.1177/2332858416673968.

Scherer, R., Siddiq, F., & Tondeur, J. (2019). The technology acceptance model (TAM): A meta-analytic structural equation modeling approach to explaining teachers' adoption of digital technology in education. *Computers & Education, 128,* 13–35.

Talamo, A., & Ligorio, B. (2001). Strategic identities in cyberspace. *Cyberpsychology Behavior, 4*(1), 109–122. doi:10.1089/10949310151088479.

Teo, T., & Zhou, M. (2016). The influence of teachers' conceptions of teaching and learning on their technology acceptance. *Interactive Learning Environments, 25*(4), 513–527. doi:10.1080/10494820.2016.1143844.

Tsuei, M., & Hsu, Y.-Y. (2019). Parents' acceptance of participation in the integration of technology into children's instruction. *The Asia-Pacific Education Researcher, 28*(5), 457–467. doi:10.1007/s40299-019-00447-3.

Vanlehn, K., Wetzel, J., Grover, S., & Sande, B. V. D. (2017). Learning how to construct models of dynamic systems: An initial evaluation of the Dragoon intelligent tutoring system. *IEEE Transactions on Learning Technologies*, *10*(2), 154–167.

Veletsianos, G., & Russell, G. S. (2014). Pedagogical agents. In *Handbook of research on educational communications and technology* (pp. 759–769), J.M. Spector et al. (eds.), Handbook of Research on Educational Communications and Technology, DOI 10.1007/978-1-4614-3185-5_61, © Springer Science & Business Media New York.

Venkatesh, V., Morris, M. G., Davis, G. B., & Davis, F. D. (2003). User acceptance of information technology: Toward a unified view. *MIS Quarterly*, *27*(3), 425–478.

Von Ahn, L. (2013). *Duolingo: Learn a language for free while helping to translate the web*. Paper presented at the 2013 International Conference on Intelligent User Interfaces.

Walton, G. M., Cohen, G. L., Cwir, D., & Spencer, S. J. (2012). Mere belonging: The power of social connections. *Journal of Personality and Social Psychology*, *102*(3), 513.

Williams, J. J., Kim, J., Rafferty, A., Maldonado, S., Gajos, K. Z., Lasecki, W. S., & Heffernan, N. (2016). *Axis*. Paper presented at the Third ACM Conference on Learning @ Scale.

Williams MD, Rana NP and Dwivedi YK (2015) The unified theory of acceptance and use of technology (UTAUT): a literature review. *Journal of Enterprise Information Management*, *28*(3), 443–488.

Woo, H. L. (2009). Designing multimedia learning environments using animated pedagogical agents: Factors and issues. *Journal of Computer Assisted Learning*, *25*(3), 203–218.

Zhu, S., Shi, Y., Wu, D., Yang, H. H., Wang, J., & Kwok, L.-F. (2014). *To be or not to be: Using tablet PCs in K-12 education*. Paper presented at the 2014 International Conference of Educational Innovation through Technology.

Złotowski, J., Proudfoot, D., Yogeeswaran, K., & Bartneck, C. (2015). Anthropomorphism: Opportunities and challenges in human–robot interaction. *International Journal of Social Robotics*, *7*(3), 347–360.

Chapter 6

The Only Living Boy in Homeroom

How Virtual Classes and Agents Fundamentally Change the Learning Experience

Andrew J. Hampton, Donald "Chip" Morrison and Brent Morgan

1 Introduction

The combination of artificially intelligent agents and increased adoption of distance learning will have inevitable impacts—both educational and social—for the next generation of learners. These impacts have barely been probed. Formal instruction has not fundamentally changed in living memory: students gather in classrooms to receive instruction from teachers, who hold knowledge that they work to impart. This generally involves some combination of giving lectures and assigning practice exercises, with known solutions against which to measure student progress (Lyle, 2008). Pedagogical advances have abounded within this framework. Most notably for our purposes, rigorous mapping of domains into discrete *knowledge components* and quantifiable *mastery* (e.g., VanLehn, 2006) has enabled standardized understanding of proficiency. This in turn enabled new kinds of standardized testing and evaluation, with accompanying policy initiatives and incentive structures, all stemming from a "correct answering" paradigm (e.g., see Morrison & Miller, 2018; Newman, Morrison, & Torzs, 1993). Those standards map well onto the structural affordances of learning management systems that can track individual and group progress at many levels of granularity. However, the essential framework of transmitting knowledge unidirectionally has stood stubbornly against challenges. This brings us to the spring of 2020, when the option for in-person instruction was forcibly removed by a global pandemic.

Though distance learning had been steadily increasing in popularity (Means, Toyama, Murphy, Bakia, & Jones, 2010), it suddenly became the only option in many schools. Ambushed instructors applied, ad hoc, conventional classroom structures, schedules, and dynamics without time to incorporate the new affordances and constraints of a natively technical medium (Ortiz, 2020; Perets et al., 2020). Critically, the

DOI: 10.4324/9781003030928-9

reduced fidelity of interactions in video conferencing has created an opportunity to leverage artificially intelligent agents, serving as both peers and instructors, to supplement more conventional approaches. Conversational interaction through these types of agents has a large body of research demonstrating potential for educational achievement in various domains (e.g., Carbonell, 1970; Johnson & Lester, 2016; Nye, Graesser, & Hu, 2014; Stevens & Collins, 1977), though generally constrained to relatively small-scale experiments and deployments. Scalable implementation of sufficiently intelligent artificial agents would allow drastically increased levels of supervised discourse learning that would not be feasible with conventional student–teacher ratios. This augmentation offers substantial advantages, and creates the opportunity for a long-romanticized phase shift from knowledge transmission to knowledge *construction*. Efforts to provide frameworks that could implement these scalable intelligent learning environments are already in progress (e.g., GIFT; Spain, Rowe, Goldberg, Pokorny, Lester, & Rockville, 2019). These could mitigate the strain caused by the difficulty of establishing synchronous class times, and reduce instructor workload by providing individualized instruction and feedback. Further, they can generate persistent learner data that are central to advanced instructional interventions based on modeling learners' behavior and individual characteristics for optimizing adaptivity.

These possibilities naturally involve massive changes not only in instructor training and intervention, but in how learners expect to receive instruction. Classrooms provide social engagement and training as well as formal educational content. Changes in language production and comprehension naturally follow changes in medium, and those observed in technologically mediated communication can prove both advantageous (e.g., lowered barriers to contribution) and deleterious (e.g., reduced empathy, increased cognitive load). Navigating those waters may prove a central challenge in establishing a successful virtual educational framework.

2 Discourse Learning

Interest in the critical role of dialogue in human learning has a long history, especially as grounded in the social learning theories that began to spread in the United States from the late 1960s—e.g., in translations of Bakhtin's ideas about "internally persuasive discourse" (reviewed in Matusov, 2007); Vygotsky's sociocultural theory (1978); Bandura's theory of social learning (1969); and, Bruner's social interactionist theory of language learning (1983). In all of these theories, a central idea is that knowledge cannot be simply "transmitted" from one mind to another, but must instead be "constructed" in the mind of the learner in the context of linguistic interactions with, as Vygotsky put it, "more knowledgeable others." Although these ideas, in various forms, came to dominate the school reform movement of the 1980s onward—such as in theories of cognitive apprenticeship (Collins, Brown, & Newman, 1988); situated learning (Lave & Wenger 1991); situated cognition (Brown, Collins, & Duguid, 1989), and more recently, "accountable talk" (Michaels, O'Connor, & Resnick, 2008)—the theoretical importance of dialogue has arguably had relatively little practical impact on classroom talk, where whole-class, teacher-led patterns of discourse continue to dominate (e.g., Lyle, 2008).

However, several new developments suggest the possibility of change. First, sociocultural and cognitive theories of learning have recently found support in evidence-based *biocultural* (Sinha, 2017) theories of language and human origins. The evidence is surfacing from a broad range of disciplines, including: evolutionary biology (Aiello & Wheeler, 1995; Hawkes, O'Connell, Jones, Alvarez, & Charnov, 1998; Locke & Bogin, 2006); evolutionary psychology (Kolodny, Edelman, & Lotem, 2015); cultural anthropology (Hewlett, Fouts, Boyette, & Hewlett, 2011; Hewlett & Roulette, 2016) archeology (Gärdenfors & Högberg, 2017); primatology (Tomasello, Call, Nagell, Olguin, & Carpenter, 1994); and neuroscience (Aboitiz, Aboitiz, & García, 2010; Schaafsma, Pfaff, Spunt, & Adolphs, 2015). Taken together, the convergent, accumulating evidence from these and many other such sources, suggests strongly that human language itself likely evolved at least partially under pressure for an enhanced signaling system that would more easily support the transmission of hard-won cultural knowledge from experts to novices, and from one generation to the next (Csibra & Gergely, 2011; Gärdenfors & Högberg, 2017; Laland, 2017; Morrison, 2018, 2020; Sterelny, 2012; Strauss & Ziv, 2012). The result is what Tomasello, Savage-Rumbaugh, & Kruger (1993) termed the "ratchet effect" of cumulative cultural transmission. Language at once supports collaborative innovation in one generation, and, through language-enabled teaching and learning, ensures that successful innovations get passed on to figure generations.

Importantly, the social construction of knowledge (defined as "partially validated propositions"; Rauch, 2018), is not just a matter of passing along chunks of authoritative knowledge and expertise from experts to novices, and from one generation to the next. Rather, human discourse—more precisely, *reasoned argumentation*—is the essential mechanism whereby propositions are created, tested, and validated through rational, evidence-based argumentation and group consensus. In this view, the whole edifice of human knowledge rests, was constructed from, and continues to be produced and refined in the crucible of productive argumentation among individual members of multiple, more or less loosely-connected groups. Although the emergence of literacy, email, and other modern communication technologies allow for asynchronous forms of debate, these developments all rest on the ancient capacity of the human brain to engage in intelligent, task-embedded discourse with other humans.

This basic socio-epistemological mechanism only works, however, if the participants in the discourse bring with them certain background knowledge, habits of mind, and specific discourse moves associated with these habits of mind. For example, "epistemic vigilance" (Sperber et al., 2010) demands that claims must be grounded in evidence (e.g., "How do you know that's a storm cloud?"), and that one must be willing to change one's mind if one's own belief is undermined by another's compelling evidence. Further, these capacities are subject to developmental (age-related) limits on what has been called "epistemological understanding" (Kuhn, Cheney, & Weinstock, 2000), and also require cultural nourishment.

3 Intelligent Tutoring Systems

This raises the question as to the potential role of modern technology in helping people (especially young people) acquire the habits of mind and discourse moves that make rational argumentation possible—in other words, a kind of "socio-epistemological engineering" (Peschl & Fundneider, 2014) through technology. Indeed, discourse-based systems have a long history of support in the educational computing community, dating back to the "Talkamatic" chat room system implemented in early (1960s-era) versions of PLATO (Woolley, 1994) and continuing through the development of AI-supported conversational agents ("intelligent tutors") capable of engaging in something, approaching natural language dialogue with learners (e.g., Carbonell, 1970; Johnson & Lester, 2016; Nye, Graesser, & Hu, 2014; Stevens & Collins, 1977). Other initiatives to build computer-supported collaborative learning environments have included CSILE ("Computer Supported Intentional Learning Environments," later "Knowledge Builder"; Bereiter & Scardamalia, 1989); Belvedere (Suthers, Toth, & Weiner, 1997); Group Scribbles (Brecht et al. 2007; Patton, Tatar, & Dimitriadis, 2008); "epistemic game" environments (e.g., Shaffer, 2006); and VIBRANT, an agent-based support system for collaborative brainstorming in scientific inquiry (Wang, Rosé, & Chang, 2011).

3.1 Conversational Agents

One-to-one (human) tutoring is often viewed as the gold standard in education. However, the inherent cost proves prohibitive to the majority of students or institutions that provide the instruction. The advancement of intelligent tutoring systems suggests one way to mitigate this problem, notably with conversational agents, though also with learner modeling, branching techniques, etc. This can offer scalable solutions that potentially reduce the per-student cost to a manageable sum (Bloom, 1984), and in turn foster a more socially equitable learning ecosystem. This provides a clear incentive to create and proliferate a broad array of artificial agents with tutoring skills commensurate with their flesh-and-blood counterparts.

Dialogue-based intelligent tutoring systems (ITS) have demonstrated success in facilitating deep learning of complex content across many domains (e.g., Graesser, 2016; Nye, Graesser, & Hu, 2014). In the basic form of these interactions, a virtual tutor (alternatively a mentor, expert, or occasionally peer) uses various limited conversational techniques to elicit complete answers and to correct misconceptions. The implementation and design vary widely, but designers commonly designate the components of an expected answer that fully address an open-ended question. Various natural language processing (NLP) techniques identify which parts are present versus absent. Expected answers, and parts of answers, may be affirmed and reiterated. For missing parts, the virtual tutor may pump for more information generically ("Can you tell me more about that?"), give hints ("What might happen with respect to D?"), or prompt by starting a statement ("If A leads to B, then C would lead to what?").

Dialogue-based approaches such as these have considerable utility. The open-ended approach forces learners to generate their own responses, which is associated with increased learning (Chi & Wylie, 2014). Additionally, the variable levels of scaffolding (i.e., pump, hint, prompt) correspond to variable assessment of mastery with modular examination of content—all yielding persistent, detailed records atypical of human tutoring. As a result, conversational tutoring provides clear diagnostic value that can inform subsequent areas of focus and methods of intervention. This proves particularly valuable in hybrid tutor systems that include multiple modes of engagement targeted to type of learning (e.g., formulaic, applied, conceptual) with a unified learning record store (Graesser et al., 2018; Hampton, 2019; Morgan, Hogan, Hampton, Lippert, & Graesser, 2020).

However, approximating professional tutors or other pedagogical interactions with agents presents myriad (and often cascading) challenges. Scripts have to offer interaction, or else the lesson does not differ substantively from a monologue lecture. However, interaction cannot occur without flexibility on both sides. Absent general artificial intelligence (which still lies beyond the horizon), flexibility necessitates anticipation of common learner responses for a single agent to adequately respond.

Learners producing the correct answer (or at least part of it) to a given question probably constitute the most common response, but the set of potential wrong answers is often substantial. This results in an unusual challenge for subject matter experts, who must imagine plausible mistakes that run counter to their training (e.g., Hestenes, Wells, & Swackhamer, 1992). Common misconceptions afford scripted responses to correct the learner, similar to the interaction that would follow a correct answer. But unexpected incorrect answers present a problem, especially with a single agent design. The options include lowering the technical threshold for an expected response (that is, the NLP biases toward classifying responses as correct to avoid rejecting correct answers in cases of uncertainty), or else ignoring the input and redirecting with a canned response (e.g., "Hmm. Can you say more about that?"). Neither leads to naturalistic conversation.

3.2 Trialogues

The inclusion of an additional *peer* virtual agent has several advantages, including more realistic interpersonal exchanges that circumvent shortcomings of NLP. For example, if the system is unsure of the learner's answer, a second agent can restate a "best guess" that fits into an existing category, directed toward the tutor agent (e.g., "Yes, I also think that [insert matched response]"). From there, the tutor can respond appropriately to that scripted approximation of the learner's statement, nominally speaking to the peer agent, but for the human learner's benefit. The tutor then has a reliable transition toward formally stating the correct answer, confirming that any misconceptions have been supplanted. This avoids providing definitively corrective feedback to an already correct answer, which can cause significant frustration.

Though programming tactics such as this provide considerable systematic improvements, the expansion of interaction dynamics proves the greater benefit. Importantly, the expansion is multiplicative, not additive. With one more virtual agent (e.g., a peer agent, in addition to the tutor agent), the possibility opens for conversations between: teacher–learner, teacher–peer, peer–learner, and teacher–learner–peer. The defined role of each agent will naturally inform the dynamics of these exchanges, which would change substantially with different assignments (e.g., imaginary friend, robot assistant). However, for the purposes of this paper, we will focus on traditional dynamics occurring within a virtual classroom.

Supplementing the teacher–learner dynamic of dialogues, the newly enabled teacher–peer exchanges constitute a method of vicarious learning (Cai, Feng, Baer, & Graesser, 2014; Graesser, Cai, Morgan, & Wang, 2017; Graesser, Li, & Forsyth, 2014). In this paradigm, the human learner absorbs information by observing interactions. The back-and-forth creates a more engaging form of information presentation, relative to monologue, while still allowing passive learning (Chi & Wylie, 2014), which may be more appropriate for hesitant or beginning students (Craig, Gholson, Brittingham, Williams, & Shubeck, 2012; Gholson et al., 2009; Millis, Forsyth, Butler, Wallace, Graesser, & Halpern, 2011). In fact, vicarious conditions have outperformed interactive conditions for low-knowledge learners across multiple studies (Craig, Sullins, Witherspoon, & Gholson, 2006; Gholson et al., 2009). Vicarious instruction can work on levels beyond basic content presentation. Moving forward, modeling proper question-asking (Craig, Gholson, Ventura & Graesser, 2000), answering behavior, and goal orientation (Twyford & Craig, 2017) can open the door to more constructive exchanges involving the human learner (i.e., training the habits of mind and discourse needed for productive exchanges). Staged arguments may encourage the learner to evaluate the relative merits of opposing sides and think critically (Graesser et al., 2017; Lehman et al., 2013).

The quality of learning while engaged in vicarious learning is greatly enhanced when including deep questions in the conversation. Deep-level reasoning questions (Graesser & Person, 1994) are associated with higher cognitive difficulty according to Bloom's (1956) taxonomy and have been shown to improve learning across a number of studies (Craig et al., 2006; 2012; Gholson et al., 2009). In addition to agent conversations, virtual classrooms can also offer multiple alternative representations of the content to improve conceptual understanding (Morgan et al., 2020). A virtual classroom could offer varying levels of interactivity depending on the learner's ability, the designer's resources, etc. For example, a classroom could offer no interactivity (full vicarious), limited interactivity (a human answers binary or multiple-choice questions), or full interactivity (i.e., an ITS).

Beyond vicarious learning while observing the two agents' conversation, the added peer–learner dimension offers several possibilities, outlined in Graesser et al. (2017). For example, learners may find themselves in a competition with a peer agent, with a tutor as the judge (Millis, Forsyth, Wallace, Graesser, & Timmins, 2017). The peer agent's answers could depend on the learner's answers, adapting as needed to keep the competition close. In another scenario, peer and tutor agents may give conflicting information, forcing the learner to determine who is right (Lehman et al., 2013). Or, for high knowledge learners, the peer can appear to be struggling and ask the learner for help, becoming a teachable agent (Graesser, Forsyth, & Lehman, 2017; Leelawong & Biswas, 2008; Millis, Forsyth, Wallace, Graesser, & Timmins, 2017; Wagster, Tan, Wu, Biwas, & Schwartz, 2007).

Though these strategies provide considerable flexibility in pedagogical approaches, restricting the interface to only two virtual agents may curtail opportunities and hinder realism. For example, a learner who performs well benefits both from instructing a struggling peer and from competing with a competent peer. Conversely, a struggling learner may benefit both from modeling ideal learning behavior of the virtual peer and by identifying with a struggling peer who asks remedial questions where the learner might have hesitated. However, having the same virtual peer playing both roles in sequential problems confounds the expectations of the learner with respect to established social dynamics and persistent theory of mind. It would be jarring if a highly competent peer suddenly has difficulty answering simple questions in the next exercise.

4 Emerging Capabilities

4.1 Multiple Virtual Agents

Just as adding a second virtual agent has a multiplicative effect (from one to four), adding a third virtual agent dramatically expands the potential social dynamics available for ITS designers to exploit. The four total interlocutors (e.g., tutor, human learner, peer 1, peer 2) have six individual interaction options (dialogues), four triadic pairings (trialogues), and a four-way conversation, totaling eleven possible modes of interaction. A fourth virtual agent expands this design space again, now reaching a formidable 26 combinations of actively engaged humans and virtual agents. This enumeration, though expansive, still only considers static arrangements—an artificial constraint employed simply to keep a manageable figure. At this point in the expansion, we argue that numerically specific nomenclature (i.e., quadrilogue and quintilogue) becomes more cumbersome than useful. We instead expand on the idea of a virtual classroom by populating it with virtual agents in addition to human learners and referring to the resulting conversational pedagogical environment as a virtual class.

These agents may be programmed to intervene individually or in tandem to create complex interactions that afford the creation of novel pedagogical strategies. These strategies expand upon existing ones for conversational ITS both linearly and combinatorially. That is, the presence of more agents affords the realistic expansion of learning strategies and rates of progression represented. It also creates possibilities for complex interactions between agents and the human learner, thereby creating a closer approximation of a(n idealized) real classroom, a scenario with which students should readily identify.

We consider the conversational design afforded by Sharable Knowledge Online (SKO; Hu, Nye, Gao, Huang, Xie, & Shubeck, 2014; Nye,

Graesser, & Hu, 2014), an online conversation-based ITS creation tool. This software program enables the inclusion of four "talking head" virtual agents simultaneously (see Figure 6.1). Each agent can play a different role. For example, a "tutor" agent may serve the same role as the original AutoTutor (Graesser, Chipman, Haynes, & Olney, 2005). In that family of systems, the tutor functions like any human tutor, providing information delivery, asking and answering questions, summarizing, and generally shepherding the learning session. A "good" student agent generally answers all but the most counter-intuitive questions correctly, and asks deep, insightful questions for the tutor to answer. The good student's questions can also advance the learning session, leading the tutor into the next topic. A "struggling" student agent will answer difficult questions incorrectly, often exhibiting common misconceptions for either the tutor or the good student to remediate. The struggling student's questions are typically shallow, though helpful for a human learner (e.g., "Which one is the independent variable again?"), or requests for repetition or elaboration (e.g., "Can you explain that more?"). The last student agent functions as a class clown, making light of the material with jokes which are seemingly unheard (and never addressed) by the other agents. Although serving as a form of shakeup agent (D'Mello, Craig, Fike, & Graesser, 2009) who does not seem to care about the lesson, the content of the jokes indicate that the agent is following the conversation

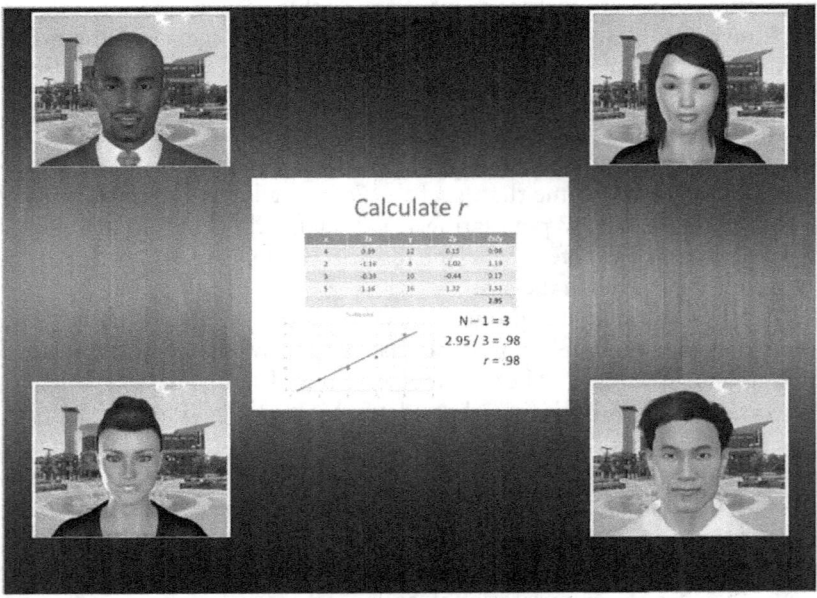

Figure 6.1 A sample of the SKO talking head overlay on learning material.

closely and has a deep understanding of the material, establishing an expectation of engagement even through irreverence.

The simplest ways in which to exploit the affordances of a virtual class involve the application of multiple strategies already discussed. As stated, learners who excel often benefit from competition with a knowledgeable peer agent. Although that same agent could offer gentle guidance (or overt criticism) to a struggling learner, the preference may be to commiserate with a peer agent of similarly low ability. Having multiple virtual agents enables transition between these designs without the need for interface alteration or awkward substitution of personas. The persistent presence of several distinct personalities allows for a kind of "greatest hits" approach to conversational ITS—peer-to-peer techniques can coexist with a tutor agent design.

Further, a change in technique from one agent to another need not exist within the same conversational frame. That is, a designer can shift the members of the virtual class by means of shifting what task the learner has to complete, creating a natural point of transition. For example, having four virtual agents always present does not necessarily afford all of the strategies described, because a peer-to-peer conversation would appear to be "monitored" by the instructor agent. Therefore, creating a scenario in which the instructor agent organically removes itself from the situation ensures closer adherence to previously validated peer-to-peer designs. Sidebars with some subset of agents create a dynamic social situation, either ad hoc or prepared, more closely aligning to in-person small group collaborative discourse. The group may split to focus on different parts of a large task, or an instructor agent may leave to eliminate an implied authority figure and foster more constructive work, or a "class clown" agent may slip a virtual note with a joke written on it. More agents creates the opportunity for more complex disagreements, with multiple coherent positions put forth and defended, all for the learner to evaluate. Shifting pedagogical approaches becomes a simple matter of changing which agent takes the lead and which agents are present—nothing is lost. This both fosters engagement through conflict and models the critical skill of productive argumentation.

4.2 Fused Interaction

To this point, we have only considered single learner scenarios, with avatars representing virtual agents. From the architecture described above, implementing ITS exercises via a conferencing platform requires only integrative (rather than innovative) engineering. One could easily imagine inserting these collaborative (i.e., a human learner with artificial agents) exercises into an online (not natively intelligent) classroom such as those adopted en masse during the COVID-19 pandemic.

From that common meeting room, human teachers would provide introductory instruction on a topic and then send students off for more interactive engagement. After the minor change of adding an avatar or virtual presence for the learners is implemented in the ITS exercises, the learner has visual representation within the virtual class that persists into breakout rooms with artificial agents. This blurs the line between the ITS scenarios detailed above and the now-common experience of an instructor managing students on a conferencing platform. In other words, the transition from full-class, instructor-led scenarios to ITS exercises become more fluid.

But perhaps the most intriguing possibilities afforded by this type of virtual class, both from pedagogical and philosophical perspectives, stem from blurring the lines between human learners and artificial agents.[1] An emphasis on avatar representation (rather than a live video feed) fosters the expectations of regular interaction with avatars in the educational environment. This policy has practical benefits indicating the likelihood of its adoption, including safety, equity, and cognitive load optimization. Avatars create a barrier to protect the privacy of minors online. They also require less bandwidth than live video feeds, giving those with less access to digital infrastructure (e.g., rural and impoverished communities) a more equal footing. And eliminating live video feeds can avoid the extraneous cognitive load of hypervigilance to social cues through monitoring and producing facial expression (Ferran & Watts, 2008). Given the import of these concerns, we may soon see the widespread adoption of avatar representation in learning environments (and potentially beyond). If that becomes a normal form of interaction, it may yield a world in which people interact with avatars without *a priori* knowledge of their status as organic or artificial intelligences.

4.3 Occasionally Artificial Peers

Theoretically, in any of the multi-agent designs above, multiple human learners could engage simultaneously with the exercise. The field of possibilities, assuming an arbitrary cap at four learners plus an instructor agent, now expands to include any combination of humans and peer agents. This ratio could vary according to constraints of the exercise (ignoring technical constraints, which we assume to be minimal given common current capabilities). For example, a difficult concept may benefit from extra scaffolding control, and therefore a more scripted, agent-based approach. Alternatively, creative exercises without predefined correct answers would naturally pair with more human-generated, unpredictable input. A balance between these two extremes would naturally shift to the latter with advancing learner age and discourse capabilities.

In any scenario beyond a fixed class size with fixed avatar representation, the distinction between human peers and artificial ones would not be immediately clear. For example, a learning system akin to the myriad examples currently in the marketplace (but not limited to a particular class) could assign learners to small groups based on their pedagogical determinations (or other factors). A large number of users precludes knowing them all personally, so learners would naturally be paired with strangers at least some of the time. The variability of human input would far outpace that of any reasonably designed set of artificial agents. The result of which, for most interactions, would be an expectation that artificial agents behave more "normal" than do humans.[2]

Alternatively, in either fixed classes or in large learning systems, a rotating avatar function could negate continuity of appearance. This would have the same effect of disguising organic versus artificial intelligence, as learners would no longer be able to recognize known peers on sight. Pedagogical and technical affordances aside, this creates an opportunity to study and potentially train avoidance of prejudice in education. Online environments such as gaming already commonly afford users the option of selecting an avatar that does not match their real-world gender, which they may do to avoid discrimination (McLean & Griffiths, 2019) or to appear to better match their behavior with gender norms (Huh & Williams, 2010). Initial implementations in the learning environment could, for example, shield a young woman who is insecure about her propensity for STEM. Broader initiatives could combat the societal reason for that insecurity by, perhaps, creating the impression of equal (or predominant) female representation in virtual STEM classes and/or explicitly confronting stereotype threat (Spencer, Steele, & Quinn, 1999). Similar initiatives could study and combat racial discrimination by both teachers and peers that may harm educational performance of African-American, Latinx, and American Indian students (e.g., Farkas, 2003).

4.4 Intermittently Artificial Instructors

Beyond peer assignment, the instructor agent could include the capacity for human teachers to assume control without notifying learners of any change. In other words, humanity is a variable. Two primary advantages would accompany such a function. First, this would approximate the common classroom structure of small-group activities, where teachers float between groups and intervene to provide instruction only as they deem necessary. A persistent chat (or closed caption) log would let the teacher catch up without interrupting conversational flow. Second, it would lend credibility to the instructor agent, who (from the learners' perspective) may actually be their human teacher at any given time. This extends beyond the Wizard of Oz experiment paradigm (Bella

&Hanington, 2012) wherein participants believe they are interacting with an artificial intelligence when it is in fact controlled by a human in real time. Instead, learners do not know from moment to moment if their instructor has organic or artificial intelligence. In such an environment, the incentive structure dictates that they should always assume the former.

5 Conclusion

Wide scale adoption of these paradigms in which human and artificial agents are not immediately distinguishable holds the potential to alter our style of interaction substantially. The incentive to treat any instructor agent as human does not necessarily transfer to peer agents. In this case, what pattern of behavior constitutes the default? If it leans toward humanity in the presence of ambiguity, will that habit extend to interactions with agents readily identifiable as artificial? Further, does increasingly polite behavior toward artificial agents reflect a simple habit, maintained as the most efficient strategy, or does it indicate that we are imbuing AI with more humanity? These are testable questions. The politeness common to interpersonal communication has distinctive, quantifiable characteristics such as "magic words" (please and thank you) and indirect speech acts (Searle, 1975). Research into the circumstances predicting level of politeness can then inform our understanding of an evolving relationship between humans and AI. With that basic model, we can investigate if patterns of behavior are sensitive to manipulations of ambiguity, perceived power differential, or purpose of the AI (e.g., an oracle, a genie; Bostrom, 2014), perceived sophistication, etc.

The strength of an ITS is its ability to adapt to the needs of the individual student. However, the ITS can adapt not just to a student's knowledge, but to *their entire way of thinking*. Advocates of neurodiversity consider several "disorders"—including attention deficit hyperactivity disorder (ADHD), autism spectrum disorder (ASD), dyslexia, and communication disorders—as merely representing the full spectrum of human mental functioning. From this perspective, learners are not "disabled," but "atypical" (Dinishak, 2016; Silberman, 2015).

ITSs already offer features which inherently cater to neurodiverse learners. Intelligent feedback and recommendations increase attention and motivation for all learners, but the system's infinite patience accommodates difficulties in focus common to those with ADHD. The system will be available when they are ready to proceed. For ASD students, dialogue-based ITSs have clearly defined conversational turns and regular structures, avoiding reliance on more nuanced social cues that can be difficult to recognize. For dyslexic students, text-to-speech and speech-to-text technology allow any dialogue to be spoken as well as typed.

Further, ITSs can offer additional adaptability specific to neurodiverse individuals. For example, for ADHD learners, it can be difficult to attend to lengthy dialogue or instructions, so adaptive segmentation might be beneficial. Similarly, a conversational ITS could offer prominent "repeat" functions and a conversation log to address attentional lapses. Motivation can also be an issue for ADHD learners, so ITSs can implement game-like features to reward interaction and correct answers. Calendaring systems that reward regular (e.g., daily) interaction can also help keep an ADHD learner plugged in. For ASD students, a conversational ITS could adapt by explicitly noting conversational markers to help the student understand why an agent said a particular comment (e.g., "Note that your peer seemed to misunderstand the core concept but the instructor provided a correction."). An ITS could also prepare feedback that rewards and encourages finely tuned attention to detail, common to ASD (e.g., "Good work reading this circuit diagram. You might enjoy this link to a complete catalogue of circuit diagram notations for more complex modeling."). For students with communication disorders, ITSs could also offer verbal input to allow students to practice verbal communication in a low-stakes environment (e.g., the infinitely patient agents do not judge them if they stutter or have delayed responses due to aphasia).

Technical advancement, pedagogical initiatives, market forces, and a global pandemic have combined to create an educational ecosystem primed for unprecedented levels of innovation. This chapter provides a partial overview of one facet of this opportunity—multiple intelligent agent designs in ITS. Yet this fraction of a sample contains the possibility of fundamentally changing our perceived relationship with AI. Learners will likely not place great emphasis on human–AI differentiation given the potential ubiquity of ambiguous interactions. This is especially true in an educational environment designed to productively tax cognitive capacity that would be wasted on such an effort. After more than 70 years of asking if AI can pass the Turing Test, that standard question of whether we can tell human from machine (e.g., Saygin, Cicekli, & Akman, 2000), how strange to think we may create an environment in which the next generation of learners does not care.

Acknowledgements

This research was supported by the National Science Foundation under the awards The Learner Data Institute (Award #1934745) and DataWhys (Award #1918751). The opinions, findings, and results are solely the authors' and do not necessarily reflect those of the National Science Foundation.

Notes

1 *Editor's note:* The essence of this approach arguably contradicts a primary conclusion of the preceding chapter by Zhang et al., who raise concerns over the long-term social effects of interacting with ambiguously human agents. We leave it to the reader to weigh the affordances presented in this chapter against those concerns, particularly within the ethical guidelines of maleficence and beneficence put forward in the previous chapter by Ness et al.
2 The absolute version of this statement assumes that the behavior of organic and artificial intelligence vary in the same way, which, of course, is not true. However, within a task-oriented conversation, variability of pre-scripted or AI-generated linguistic behavior is significantly constrained and less likely to be recognized as non-human.

References

Aboitiz, F., Aboitiz, S., & García, R. R. (2010). The phonological loop: a key innovation in human evolution. *Current Anthropology, 51*(S1), S55–S65.

Aiello, L. C., & Wheeler, P. (1995). The expensive-tissue hypothesis: The brain and the digestive system in human and primate evolution. *Current Anthropology, 36*(2), 199–221.

Bella, M., & Hanington, B. (2012). *Universal Methods of Design*. Beverly, MA: Rockport Publishers.

Scardamalia, M., Bereiter, C., McLean, R. S., Swallow, J., & Woodruff, E. (1989). Computer-supported intentional learning environments. *Journal of Educational Computing Research, 5*(1), 51–68.

Bloom, B. (1956). *A Taxonomy of Cognitive Objectives*. New York: McKay.

Bloom, B. S. (1984). The 2 sigma problem: The search for methods of group instruction as effective as one-to-one tutoring. *Educational Researcher, 13*(6), 4–16.

Bostrom, N. (2014). *Superintelligence: Paths, Dangers, Strategies*. Oxford, UK: Oxford University Press.

Brecht, J., DiGiano, C., Patton, C., Tatar, D., Chaudhury, S. R., Roschelle, J., & Davis, K. (2007). Coordinating networked learning activities with a general-purpose interface. In *Proceedings of the 5th World Conference on Mobile Learning*. Retrieved from https://iamlearn.org/wp-content/uploads/2018/01/mLearn2006_Proceedings.pdf

Brown, J. S., Collins, A., & Duguid, P. (1989). Situated cognition and the culture of learning. *Educational Researcher, 18*(1), 32–42.

Cai, Z., Feng, S., Baer, W., & Graesser, A. (2014). Instructional strategies in trialogue-based intelligent tutoring systems. In R. Sottilare, A.C. Graesser, X. Hu, and B. Goldberg (Eds.), *Design Recommendations for Intelligent Tutoring Systems: Instructional Management, Volume 2* (pp. 225–235). Orlando, FL: Army Research Laboratory.

Carbonell, J. R. (1970). AI in CAI: An artificial-intelligence approach to computer-assisted instruction. *IEEE Transactions on Man–Machine Systems, 11*(4), 190–202.

Chi, M. T., & Wylie, R. (2014). The ICAP framework: Linking cognitive engagement to active learning outcomes. *Educational Psychologist, 49*(4), 219–243.

Collins, A., Brown, J. S., & Newman, S. E. (2018). *Cognitive Apprenticeship: Teaching the Crafts of Reading, Writing, and Mathematics* (pp. 453–494). London: Routledge.

Craig, S. D., Gholson, B., Brittingham, J. K., Williams, J. L., & Shubeck, K. T. (2012). Promoting vicarious learning of physics using deep questions with explanations. *Computers & Education, 58*(4), 1042–1048.

Craig, S. D., Gholson, B., Ventura, M., & Graesser, A. (2000). The Tutoring Research Group: Overhearing dialogues and monologues in virtual tutoring sessions. *International Journal of Artificial Intelligence in Education, 11,* 242–253.

Craig, S. D., Sullins, J., Witherspoon, A., & Gholson, B. (2006). The deep-level-reasoning-question effect: The role of dialogue and deep-level-reasoning questions during vicarious learning. *Cognition and Instruction, 24*(4), 565–591.

Csibra, G., & Gergely, G. (2011). Natural pedagogy as evolutionary adaptation. *Philosophical Transactions of the Royal Society B: Biological Sciences, 366*(1567), 1149–1157.

Dinishak, J. (2016). The deficit view and its critics. *Disability Studies Quarterly, 36*(4), 5.

D'Mello, S., Craig, S., Fike, K., & Graesser, A. (2009, July). Responding to learners' cognitive-affective states with supportive and shakeup dialogues. In *International Conference on Human-Computer Interaction* (pp. 595–604). Berlin, Heidelberg: Springer.

Farkas, G. (2003). Racial disparities and discrimination in education: What do we know, how do we know it, and what do we need to know? *Teachers College Record, 105*(6), 1119–1146.

Ferran, C., & Watts, S. (2008). Videoconferencing in the field: A heuristic processing model. *Management Science, 54*(9), 1565–1578.

Gärdenfors, P., & Högberg, A. (2017). The archaeology of teaching and the evolution of Homo docens. *Current Anthropology, 58*(2), 188–208.

Gholson, B., Witherspoon, A., Morgan, B., Brittingham, J. K., Coles, R., Graesser, A. C., Sullins, J., & Craig, S. D. (2009). Exploring the deep-level reasoning questions effect during vicarious learning among eighth to eleventh graders in the domains of computer literacy and Newtonian physics. *Instructional Science, 37*(5), 487–493.

Graesser, A. C. (2016). Conversations with AutoTutor help students learn. *International Journal of Artificial Intelligence in Education, 26,* 124–132.

Graesser, A. C., Cai, Z., Morgan, B., & Wang, L. (2017). Assessment with computer agents that engage in conversational dialogues and trialogues with learners. *Computers in Human Behavior, 76,* 607–616.

Graesser, A. C., Chipman, P., Haynes, B. C., & Olney, A. (2005). AutoTutor: An intelligent tutoring system with mixed-initiative dialogue. *IEEE Transactions on Education, 48*(4), 612–618.

Graesser, A. C., Forsyth, C. M., & Lehman, B. A. (2017). Two heads may be better than one: Learning from computer agents in conversational trialogues. *Teachers College Record, 119*(3), 1–20.

Graesser, A. C., Hu, X., Nye, B. D., VanLehn, K., Kumar, R., Heffernan, C., Heffernan, N., Woolf, B., Olney, A. M., Rus, V., Andraskik, F., Pavlik, P., Cai, Z., Wetzel, J., Morgan, B., Hampton, A. J., Lippert, A. M., Wang, L.,

Cheng, Q., Vinsen, J. E., Kelly, C. N., McGlown, C., Majmudar, C. A., Morshed, B., & Baer, W. (2018). ElectronixTutor: An intelligent tutoring system with multiple learning resources. *International Journal of STEM Education, 5*(15), 1–21.

Graesser, A. C., & Person, N. K. (1994). Question asking during tutoring. *American educational Research Journal, 31*(1), 104–137.

Hampton, A. J. (2019). Conversational AIS as the cornerstone of hybrid tutors. In R. A. Sottilare & J. Schwarz (Eds.) *Proceedings of the First International Conference, AIS 2019, Held as Part of the 21st HCI International Conference* (634–644), Orlando, FL, USA, July 26–31, 2019.

Hawkes, K., O'Connell, J. F., Jones, N. B., Alvarez, H., & Charnov, E. L. (1998). Grandmothering, menopause, and the evolution of human life histories. *Proceedings of the National Academy of Sciences, 95*(3), 1336–1339.

Hestenes, D., Wells, M., & Swackhamer, G. (1992). Force concept inventory. *The Physics Teacher, 30*(3), 141–158.

Hewlett, B. S., Fouts, H. N., Boyette, A. H., & Hewlett, B. L. (2011). Social learning among Congo Basin hunter–gatherers. *Philosophical Transactions of the Royal Society B: Biological Sciences, 366*(1567), 1168–1178.

Hewlett, B. S., & Roulette, C. J. (2016). Teaching in hunter–gatherer infancy. *Royal Society Open Science, 3*(1), 150403.

Hu, X., Nye, B. D., Gao, C., Huang, X., Xie, J., & Shubeck, K. (2014, June). Semantic representation analysis: A general framework for individualized, domain-specific and context-sensitive semantic processing. In *International Conference on Augmented Cognition* (pp. 35–46). Cham: Springer.

Huh, S., & Williams, D. (2010). Dude looks like a lady: Gender swapping in an online game. In W. S. Bainbridge (Ed.) *Online Worlds: Convergence of the Real and the Virtual* (pp. 161–174). London: Springer.

Johnson, W. L., & Lester, J. C. (2016). Face-to-face interaction with pedagogical agents, twenty years later. *International Journal of Artificial Intelligence in Education, 26*(1), 25–36.

Kolodny, O., Lotem, A., & Edelman, S. (2015). Learning a generative probabilistic grammar of experience: A process-level model of language acquisition. *Cognitive Science, 39*(2), 227–267.

Kuhn, D., Cheney, R., & Weinstock, M. (2000). The development of epistemological understanding. *Cognitive Development, 15*(3), 309–328.

Laland, K. N. (2017). The origins of language in teaching. *Psychonomic Bulletin & Review, 24*(1), 225–231.

Lave, J., & Wenger, E. (1991). *Situated Learning: Legitimate Peripheral Participation.* Cambridge: Cambridge University Press.

Leelawong, K., & Biswas, G. (2008). Designing learning by teaching agents: The Betty's Brain system. *International Journal of Artificial Intelligence in Education, 18*(3), 181–208.

Lehman, B., D'Mello, S., Strain, A., Mills, C., Gross, M., Dobbins, A., Wallace, P., Millis, K, & Graesser, A. (2013). Inducing and tracking confusion with contradictions during complex learning. *International Journal of Artificial Intelligence in Education, 22*(1–2), 85–105.

Locke, J. L., & Bogin, B. (2006). Language and life history: A new perspective on the development and evolution of human language. *Behavioral and Brain Sciences, 29*(3), 259–280.

Lyle, S. (2008). Dialogic teaching: Discussing theoretical contexts and reviewing evidence from classroom practice. *Language and Education, 22*(3), 222–240.

Matusov, E. (2007). Applying Bakhtin scholarship on discourse in education: A critical review essay. *Educational Theory, 57*(2), 215–237.

McLean, L., & Griffiths, M. D. (2019). Female gamers' experience of online harassment and social support in online gaming: A qualitative study. *International Journal of Mental Health and Addiction, 17*(4), 970–994.

Means, B., Toyama, Y., Murphy, R., Bakia, M., & Jones, K. (2010). *Evaluation of Evidence Based Practices in Online Learning: A Meta Analysis and Review of Online Learning Studies.* Washington, DC: U.S. Department of Education.

Michaels, S., O'Connor, C., & Resnick, L. B. (2008). Deliberative discourse idealized and realized: Accountable talk in the classroom and in civic life. *Studies in Philosophy and Education, 27*(4), 283–297.

Millis, K., Forsyth, C., Butler, H., Wallace, P., Graesser, A., & Halpern, D. (2011). Operation ARIES!: A serious game for teaching scientific inquiry. In M. Ma, A. Oikonomou, & L. Jain (Eds.) *Serious Games and Edutainment Applications* (pp. 169–195). London: Springer.

Millis, K., Forsyth, C., Wallace, P., Graesser, A. C., & Timmins, G. (2017). The impact of game-like features on learning from an intelligent tutoring system. *Technology, Knowledge and Learning, 22*(1), 1–22.

Morgan, B., Hogan, A. M., Hampton, D., Lippert, A., & Graesser, A. C. (2020). The need for personalized learning and the potential of intelligent tutoring systems. In P. Kendeou, P. Van Meter, D. Lombardi, & A. List (Eds.) *Handbook of Learning from Multiple Representations and Perspectives.* New York: Routledge.

Morrison, D. M., & Miller, K. B. (2018). Teaching and learning in the pleistocene: A biocultural account of human pedagogy and its implications for AIED. *International Journal of Artificial Intelligence in Education, 28*(3), 439–469.

Newman, D., Morrison, D., & Torzs, F. (1993). The conflict between teaching and scientific sense-making: The case of a curriculum on seasonal change. *Interactive Learning Environments, 3*, 1–16.

Nye, B. D., Graesser, A. C., & Hu, X. (2014). AutoTutor and family: A review of 17 years of natural language tutoring. *International Journal of Artificial Intelligence in Education, 24*(4), 427–469.

Ortiz, P. (2020). Teaching in the time of COVID-19. *Biochemistry and Molecular Biology Education.* doi: 10.1002/bmb.21348.

Patton, C. M., Tatar, D., & Dimitriadis, Y. (2008). Trace Theory, Coordination Games, and Group Scribbles. In J. Voogt (Ed.) *International Handbook of Information Technology in Primary and Secondary Education* (pp. 921–933). Boston, MA: Springer.

Perets, E. A., Chabeda, D., Gong, A. Z., Huang, X., Fung, T. S., Ng, K. Y., Bathgate, M., & Yan, E. C. (2020). Impact of the emergency transition to remote teaching on student engagement in a non-STEM undergraduate chemistry course in the time of COVID-19. *Journal of Chemical Education, 97*(9), 2439–2447.

Peschl, M. F., & Fundneider, T. (2014). Designing and enabling spaces for collaborative knowledge creation and innovation: From managing to enabling innovation as socio-epistemological technology. *Computers in Human Behavior, 37,* 346–359.

Rauch, J. (2018). The constitution of knowledge. *National Affairs, 37*(4), 125–137.

Saygin, A. P., Cicekli, I., & Akman, V. (2000). Turing test: 50 years later. *Minds and Machines, 10*(4), 463–518.

Schaafsma, S. M., Pfaff, D. W., Spunt, R. P., & Adolphs, R. (2015). Deconstructing and reconstructing theory of mind. *Trends in Cognitive Sciences, 19*(2), 65–72.

Searle, J. R. (1975). Indirect speech acts. In P. Cole & J. L. Morgan (Eds.) *Syntax and Semantics: Volume 3—Speech Acts* (pp. 59–82). Leiden: Brill.

Shaffer, D. W. (2006). Epistemic frames for epistemic games. *Computers & Education, 46*(3), 223–234.

Silberman, S. (2015). *Neurotribes: The Legacy of Autism and the Future of Neurodiversity.* London: Penguin.

Sinha, C. (2017). *Ten Lectures on Language, Culture and Mind.* Leiden: Brill.

Spain, R., Rowe, J., Goldberg, B., Pokorny, R., Lester, J., & Rockville, M. D. (2019). Enhancing learning outcomes through adaptive remediation with GIFT. In *Proceedings of the Interservice/Industry Training, Simulation and Education Conference* (pp. 1–11).

Spencer, S. J., Steele, C. M., & Quinn, D. M. (1999). Stereotype threat and women's math performance. *Journal of Experimental Social Psychology, 35*(1), 4–28.

Sperber, D., Clément, F., Heintz, C., Mascaro, O., Mercier, H., Origgi, G., & Wilson, D. (2010). Epistemic vigilance. *Mind & Language, 25*(4), 359–393.

Sterelny, K. (2012). *The Evolved Apprentice.* Cambridge, MA: MIT press.

Stevens, A. L., & Collins, A. (1977). The goal structure of a Socratic tutor. In *Proceedings of the National ACM Conference* (pp. 256–263). New York: Association of Computing Machinery.

Strauss, S., & Ziv, M. (2012). Teaching is a natural cognitive ability for humans. *Mind, Brain, and Education, 6*(4), 186–196.

Suthers, D. D., Toth, E., & Weiner, A. (1997). An integrated approach to implementing collaborative inquiry in the classroom. In R. Hall, N. Miyake, & N. Enyedy (Eds.) *Proceedings of the 2nd International Conference on Computer Support for Collaborative Learning* (Vol. 97, pp. 272–279). Toronto, ON: International Society of the Learning Sciences.

Tomasello, M., Call, J., Nagell, K., Olguin, R., & Carpenter, M. (1994). The learning and use of gestural signals by young chimpanzees: A transgenerational study. *Primates, 35*(2), 137–154.

Tomasello, M., Savage-Rumbaugh, S., & Kruger, A. (1993). Imitative learning of actions on objects by children, chimpanzees and enculturated chimpanzees. *Child Development, 64,* 1688–1705.

Twyford, J., & Craig, S. D. (2017). Modeling goal setting within a multimedia environment on complex physics content. *Journal of Educational Computing Research, 55*(3), 374–394.

Van Lehn, K. (2006). The behavior of tutoring systems. *International Journal of Artificial Intelligence in Education, 16*(3), 227–265.

Wagster, J., Tan, J., Wu, Y., Biwas, G., & Schwartz, D. (2007). Do learning by teaching environments with metacognitive support help students develop better learning behaviors?. In D. S. McNamara & J. G. Trafton (Eds.), *Proceedings of the 29th Annual Cognitive Science Society* (pp. 695–700). Austin, TX: Cognitive Science Society.

Wang, H. C., Rosé, C. P., & Chang, C. Y. (2011). Agent-based dynamic support for learning from collaborative brainstorming in scientific inquiry. *International Journal of Computer-Supported Collaborative Learning, 6*(3), 371–395.

Woolley, D. R. (1994). PLATO: The emergence of online community. In J. Malloy (Ed.) *Social Media Archeology and Poetics.* Cambridge, MA: MIT Press.

Part 3

Decisions, Decisions

Chapter 7

J.A.R.V.I.S., What Should I Do Now? Human Virtual Agents as a Means to More Ethical Decision-Making

Jeanine A. DeFalco and John Hart

1 Introduction

Popular fiction comfortably and repeatedly relies on virtual agents to help humans make good decisions. The functionality of these virtual agents often includes the capacity to undermine decisions of unreasonable military bosses, as seen in the supercomputer "Karl" of Arthur C. Clarke's short story *The Pacifist* (1956), as well as provide advice and save lives as seen in J.A.R.V.I.S. (Just A Rather Very Intelligent System), the ultimate virtual butler from the *Iron Man* films (beginning with Arad & Feige, 2008). As is the case in much of the history of technological innovation, fiction continues to inform the scope and aims of AI-driven products. However, we believe that while our engineers and designers continue to strive to turn the fiction of yesterday into the commonplace of today, it is of critical importance that we encourage both the developers and consumers of these innovations to consider how to design and employ these systems responsibly and ethically. This chapter will review the history of virtual humans, discuss how these systems can be used in supporting the ethical decision-making capabilities of military personnel in particular, and provide some initial thoughts on how to begin designing and integrating ethical measurements into virtual agent recommender and instructional systems.

2 Virtual Human History

Over the years, human virtual agents have taken on many forms and roles over a range of domains that include education, medicine, health/mental care, and entertainment. The idea of a computer agent holding a conversation goes back to the 1950s and the ideas of Turing (1950). In the 1960s, Weizenbaum created ELIZA. Though ELIZA did not have a visual human embodiment, this was an early example of an agent successfully allowing humans to interact socially with a software agent (1966). A more modern example of a successful human virtual agent is

IBM's Watson. In 2011, Watson competed and beat two top-level human contestants on the popular trivia show, *Jeopardy!* (Markoff, 2011). Today virtual human agents such as Siri, Alexa, and Google Assistant allow us to interact with computers on a social level approaching interactions once solidly in the realm of science fiction.

Many contemporary popular agents operate exclusively in the aural domain, but some have incorporated a visual representation. These agents may allow people to interact through both verbal and nonverbal channels. One such agent was *Clippy*, Microsoft's anthropomorphic paperclip who tried to help people but generally only succeeded in annoying them (Rudman & Zajicek, 2006). Some companies have found success in using these virtual human agents, like IKEA (Anna) and Alaska Airlines (Jenn) (Chattaraman, Kwon, & Gilbert, 2012). The US Army has its own virtual sergeant (SGT Star), who even made appearances as a life-sized virtual human to answer questions about the Army and Army life (Artstein et al., 2009). Each one of these organizations allows people to interact with each agent through their websites. These embodied agents have been called avatars, virtual humans, and embodied conversational agents.

Though each term's definition includes the visual representation of a human (or humanoid character), there are some generally accepted differences (Table 7.1). An avatar is the character representation of an individual in a virtual or game environment and is controlled by that individual (Von Der Pütten et al., 2010). A virtual human is a software agent that looks, acts, and responds so as to approximate a human. In some cases, an advanced agent can even have goals as they interact with humans and other agents within their environment (Gratch et al., 2002). The difference being that the virtual human can act on its own without the control of an individual.

Table 7.1 Terms used to describe virtual agents with a visual appearance

Term	Inputs	Agent Response (Outputs)	Control of Agent Behaviors
Avatar	Game controller or keyboard	Audio	Human
Chatbot	Text or voice	Primarily text; some text to speech	AI; Rules; Scripts
Embodied conversational agent	Primarily voice; can use text	Audio	AI; Rules; Scripts
Virtual human	Primarily voice; can use text	Audio	AI; Rules; Scripts

The terms *embodied conversational agent* and *virtual human* are almost interchangeable. The primary difference is in the role that the agent assumes in the interaction. The embedded conversational agent assumes the role of a learner in a typical face-to-face dialogue where a virtual human assumes the role of a virtual actor or role-player talking to the learner (Hart & Proctor, 2019).

3 Using Surrogate Virtual Agents in the Army

Virtual agents now play many different roles in people's daily lives. In the case of agents like Alexa, Siri, and Google Assistant, they provide information like news, weather, sports scores, and even answers to trivia questions. They also play a role in assisting with tasks like seeing who is at the front door and organizing calendars. These disembodied chatbots play the role of our virtual assistants in order to help make our lives easier or better. Other agents that function as performance assistance roles are embodied agents like Anna (IKEA) and Jenn (Alaska Airlines).

Virtual agents have also found success in the training and education areas as tutors and instructors. Researchers have found that interacting with an embodied pedagogical agent increases student motivation, engagement, and even learning transfer (Baylor, 2011). Other agents in education and learning have taken on a more active role in the training process by acting as a virtual human role-player. As a role-player, the agent is not standing off to the side but is engaged in the learning process as a partner in the training scenarios. For example, Ada and Grace are virtual twins who answer questions from museum visitors on aspects of computer science and virtual humans (Swartout et al. 2010). Where the twins interact with a larger audience, other virtual humans interact via intimate social dialog. Tartaro and Cassell's Authorable Virtual Peer supports social skills development for children with Autism Spectrum Disorder (Tartaro et al., 2014).

Other virtual humans assist in overcoming fear of public speaking (Chollet et al., 2015), developing job interview skills (Baur et al., 2013), and even with medical interview or counseling skills (Johnsen et al., 2007; Rizzo et al., 2016). The military has found success in training language and cultural skills (Hill et al., 2006; Johnson & Lewis, 2015; Johnson, Lewis, & Valente, 2008) as well as developing leadership skills (Campbell et al., 2011; Hart & Proctor, 2019). The value in the use of virtual humans in the training and education process is that these agents do not tire, do not vary their performance, and impart no moral judgment on the student.

The military has used simulators in its training for decades. Some of the earliest use of simulators for training dates back to the "Blue Box" flight simulator developed by Ed Link. The "Blue Box" was used to train pilots how to fly using their instruments. Since then, trainers have developed and the military has used simulators to train all sorts of tasks needed to ensure that soldiers, sailors, marines, and airmen are able to successfully accomplish their mission. The latest area for military training is in social interactions. Virtual humans are the "flight simulators" for human face-to-face interactions. They allow trainees to practice conversations that may be difficult such as casualty notifications or dealing with a victim of sexual assault. Other areas where a virtual human can be utilized are in leadership training.

Leadership is fundamental to the Army and other military services. Leaders need to be developed with the principles that allow the Army to accomplish its mission. One particular area of interest to the Army is counseling. Surveys have identified that dealing with problems such as performance, discipline, and other personal issues are interpersonal skills that need to be developed. These surveys have also pointed to the need for training that provides hands-on experiences (Hatfield et al., 2011).

Some examples of simulations that focus on social interactions developed by the Army include Emergent Leader Immersive Training Environment (ELITE; Wansbury, Hill, & Belman, 2014), The Bilateral Negotiation (BiLAT) simulation (Hill et al., 2006), the Tactical Questioning simulation (Traum et al., 2007), and the Tactical Language Training (Yates, 2011). These systems use virtual agents to interact with human participants in similar ways to how one would interact with live role-players. Major advantages to these types of simulated interactions are that there is no associated cost to hire the role-players and the computer characters interact with consistent behaviors in every interaction (Johnsen, 2008; Kenny et al., 2007; Swartout, 2010). The disadvantages center on technological limitations of interfaces, hardware, and technologies such as speech recognition, natural language processing, and limited domain knowledge of computer characters (Hart & Proctor, 2016; Traum, 2008).

A potential new area for learning through simulation is interpersonal skills training. Large swaths of the economy involve some level of customer interaction with diverse populations, like healthcare and hospitality. Another example, the military is developing requirements for interpersonal skills training as a part of Counter Insurgency and Stability Operations, that is, missions that require them to operate amongst populations with diverse religious, ethnic, cultural, and societal values (Department of the Army, 2008; FM 3–07 Stability, 2014). For these operations to be successful, military leaders need to interact with the local leaders and populations in ways that ultimately build trust (Jones & Muñoz, 2010; Kilcullen, 2009).

Still another area for learning through simulation mediated by human virtual agents is leadership skills, particularly as they relate to developing ethical decision-making that can assist in preparing for future battlefields—cyber, urban, subterranean—where conflict zones are hallmarked by chaos, disrupted communication, and unexpected enemy combatants. This is an area of particular interest in the domain of simulation and training because the chaos of battle forces decision making in novel and hostile circumstances, requiring quick judgments without time to deliberate. Our military personnel would be well served to have opportunities to integrate dilemmas that incorporate ethical decision making ambiguities in their training prior to deployment.

4 Ethics and the Military

"Ethics" comes from the Greek word *ethos*, or character. The *Nicomachean Ethics* of Aristotle notes that ethics is principally concerned with actions oriented toward the greater good and general welfare of a community (Reed et al., 2016). The chief goal of ethics education is to develop the necessary skills for ethical decision making, which includes moral sensitivity and moral reasoning (Park et al., 2012). Increasingly over the past few decades, ethics education has become a topical preoccupation with armed forces worldwide, and its focus has included ethics training to ensure ethical leadership and decision-making at all levels beginning with entry-level military members (Emonet, 2020).

In 2006, the annual theatre-wide Mental Health Assessment Team (MHAT) survey of the wellbeing of US soldiers deployed in Iraq included ethical issues for the first-time. Specifically, the survey included issues of battlefield ethics and the efficacy of battlefield ethics training for soldiers preparing for combat operations in Iraq. The results indicated that less than 50% of the soldiers were willing to report ethical violations of members of their units. And though there was not a controlled statistical analysis of related factors, soldiers who either met criteria for PTSD or who had high levels of combat exposure reported high levels of unethical conduct (Warner et al., 2011). This led to the recommendation for battlefield ethics training and prompted the Commanding General of the Multi-National Division-Center in Iraq to create and deploy battlefield ethics training programs to soldiers under his command.

A training development team, including members of Division Psychiatrists with representatives for the Staff Judge Advocate (legal), Division Surgeon (medical), Adjutant (personnel), and Chaplain, produced a report that established the importance of ethical behavior on the battlefield, the effect of violations, and the influence of groupthink on decision making in ethical dilemmas. In addition, the team asserted the need for systematic, continuous training, focusing on instilling particular values and enhancing moral reasoning and judgment. What followed was ethical training for all brigades under the Multi-National Division Center,

Iraq (about 3,500 Soldiers) and included a leader-led, interactive program using video vignettes to highlight learning points and to use as prompts for dialogic engagement (Warner et al., 2011). The results of this work included a decreased rate of unethical conduct in all categories after training, and that PTSD was not associated with unethical conduct after controlling for combat experience. However, further studies were needed to refine the battlefield ethics training methods—including sustainability and reproducibility of training implementation (Warner et al., 2011). It is within this context that there is an identified need to determine whether personalized ethical training and sustained support can be implemented and delivered through the most current, innovative technology available, such as the development of a responsive, human-virtual agent companion.

5 Effects of Ethical Dilemmas, Moral Injury, and Stress on Combat Readiness and Recovery

It is important to note that although the Warner et al. (2011) work did not find PTSD to be associated with unethical conduct, the study was not randomized nor conducted as a controlled experimental design. In addition, it is important to distinguish that the lack of correlation of PTSD in regard to unethical conduct does not mean that the reverse is not true: that unethical conduct can have an effect on emerging degradation of mental health and the etiology of PTSD. Indeed, Brigadier General McMaster anecdotally notes that combat trauma can often be linked back to a lack of effective leadership, including unethical behavior. This position is bolstered by more recent empirical evidence supporting the interaction of engagement in unethical conduct, or morally injurious experiences (MIEs), and the etiology of health problems (Currier, Holland, & Malott, 2015).

Currier, Holland, and Malott (2015) examined the interplay between MIEs and how these experiences factor in the etiology of mental health problems of veterans. MIEs can include acts of commission or omission, harmful acts perpetrated by peers, witnessing suffering and the effects of violence and injustice during deployment. They are essentially rooted in experiences that cannot be justified within a member's personal and moral beliefs (Brémault-Phillips, et al., 2019; Currier, Holland, & Malott, 2015; Molendijk, 2019). MIEs can also occur without imminent threat of physical injury or death (e.g., witnessing the aftermath of violence; betrayal by a commanding officer or trusted civilian) as well as in the face of life-threatening stressors that have been identified as the basis for PTSD (Currier, Holland, & Malott, 2015; Molendijk, 2019).

The interplay between MIEs and mental health is further complicated when factoring in the significant effect that stress and stressful situations have on ethical behavior beyond mere recognition of ethical dilemmas

(Selart & Johansen, 2011). There is ample evidence of impairment on the executive functions of working memory and cognitive flexibility as a result of acute stress (i.e., recent, transient occurrence of a single stressor as opposed to chronic, sustained stress) (Shields, Sazma, & Yonelinas, 2016). Though acute stress is endemic in combat engagements, it is an unpredictable factor that can best be mitigated through thorough training and preparation of the warfighter. In a similar way, resistance and recovery from MIEs can be addressed through re-establishing a stable framework of beliefs, values, and moral codes, and by employing sensemaking strategies in military critical incidents (de Graaff et al., 2019).

Sensemaking is the cognitive process of determining the nature of one's circumstances and the appropriate response when facing ambiguous circumstances or novel, challenging situations (Weick, Sutcliffe, & Obstfeld, 2005). Sensemaking tactics are tied to one's ability to make quick decisions drawn from previous experiences (for example, relying on one's working memory of established mental models as developed through ethical education and training), an ability to predict consequences (cognitive flexibility), and assistance from others to recognize and interpret situations (dialogic meaning making) (de Graaff et al., 2019). Sensemaking is particularly important in stressful, chaotic circumstances, e.g., fire crews (Landgren, 2005) and the military (Kramer, van Bezooijen, & Delahaij, 2010), where units face unanticipated, critical incidents in which rapid response and accuracy leave little time for thorough consideration and deliberation (de Graaff et al., 2019).

Supporting sensemaking and, more broadly, ethical education and training requires establishing a practical ethics education based on effective pedagogical tools such as case studies that can provide a framework for reflection and analysis (Emonet, 2020; Micewski, 2016) and embedding ethics education into all training tasks (Robinson, De Lee, & Carrick, 2008). But this is not sufficient in the most critical scenarios. To supplement these, organizations such as the military would be well served to determine whether this training and education can be realized in new educational delivery platforms, particularly systems such as adaptive instructional systems, recommender systems, and virtual human companions. For the purposes of this paper and our intended research plan, however, we limit discussion to exploring the effect of a human virtual agent on the development of ethical education, particularly in the area of moral sensitivity and awareness.

6 Ethics and Decision Making: Relevant Cognitive Skills

To address the development of ethical decision making, ethics education and training should begin with an understanding of the cognitive skills and individual traits that influence how people represent and frame a problem (Simkins & Steinkuehler, 2008). To accomplish this, further investigations are necessary to unpack the factors that foster cognitive flexibility and how individuals represent information—elements related to creative and analogical thinking (DeFalco & Sinatra, 2019; Kavathatzopoulos, 1993). In addition, there is evidence that personality traits—specifically those that score low on the Honesty–Humility trait as measured by the HEXACO-PI—are predictors of harmful and unethical behaviors (Marcus & Roy, 2019), and lower learning outcomes in medical critical care education (DeFalco & Sinatra, 2019). Accordingly, there is an argument to orient research and development of human virtual systems by first unpacking the correlational relationships between an individual's moral sensitivity or awareness, their cognitive flexibility as determined through creative/analogical thinking capabilities, and personality traits. To this end, we next explore measures of these constructs and their background in relating to key concepts in ethical decision making.

6.1 Personality Traits: HEXACO

Personality traits are related to creativity (Johnson, Lewis, & Valente, 2008). In a review of the literature, Batey and Furnham (Batey & Furnham, 2006) note that while creativity in terms of the production of ideas is related to intelligence, creativity as originality rests largely on personality factors. Earlier research in this area focused on Eysenck's Gigantic Three orthogonal dimensions or superfactors of personality (Eysenck & Eysenck, 1975) and was followed by research on the Big Five Factor (Chamorro-Premuzic & Furnham, 2008; Martindale & Dailey, 1996).

For example, within the Big Five, "openness to experience" correlates with scientific and artistic creativity, divergent thinking, imagination, and originality (Erdheim, Wang, & Zickar, 2006), as well as deep and surface learning (Chamorro-Premuzic & Furnham, 2008). Perhaps most interestingly, openness has been closely associated with spatial allocation of attention—the way in which attention is directed to enhance cognitive processing (Wilson et al., 2016).

However, a more recent instrument has been substituting the Big Five: The HEXACO-PI (Ashton & Lee, 2007). The value of this instrument rests both in the history of its development (constructed from lexical studies of personality structures in diverse languages), and the introduction of a sixth factor: *Honesty-Humility*, defined as sincere, fair, modest, and unassuming versus sly, deceitful, greedy, and pretentious. As such, the HEXACO-PI model evaluates an individual's personality traits along the following criteria: Honesty–Humility (H), Emotionality (E), Extraversion (X), Agreeableness (A), Conscientiousness (C), and Openness to Experience (O) (Ashton et al., 2004).

The advantages using the HEXACO-PI scale, and the subsequently validated shorter versions HEXACO-60, over measures of the Big Five include the following:

1. The inclusion of the Honesty–Humility scale splits the five-factor agreeableness into two traits: Honesty–Humility and Agreeableness (Silvia, Martin, & Nusbaum, 2009).
2. The HEXACO–60 scales of Honesty–Humility, Agreeableness (versus Anger), and Emotionality assess dimensions that are interpreted parsimoniously in terms of constructs of reciprocal and kin altruism that are absent in the Big Five Agreeableness and Neuroticism dimensions (Ashton et al., 2004).
3. The HEXACO–60 scales correspond to the largest set of personality dimensions that is obtained from the indigenous personality lexicons of diverse human languages, where the scales of Big Five instruments are not so aligned (Ashton et al., 2004).

But perhaps the most compelling reason to the HEXACO-60 instrument is found in the research by Tybur and de Vries (2013) in which the honesty–humility trait accounted for unique variance in sensitivities to moral disgust, and Ashton and Lee's work (2007; 2009; Ashton et al., 2004) demonstrating a negative relationship between the honesty–humility trait and moral violations in workplace delinquency.

6.2 Moral Judgment: Defining Issues Test (DIT)

The Defining Issues Test (DIT) has a 35-year history as an instrument that has demonstrable efficacy in moral judgement research (Rest et al., 1999). It is rooted in the Kohlbergian model of moral development, and measures the default schema by which individuals interpret moral issues (Thoma & Dong, 2014). This measure is effective at assessing the shift from a conventional/maintaining norms perspective to a post conventional view of social cooperation (Narvaez & Bock, 2002; Thoma & Dong, 2014). Designed to capture moral schema changes that typically occur throughout adolescence to early adulthood—known as the "discovery of society" stage—the DIT captures a person's understanding of the relationship of the individual to institutions, established practices, and systems of social norms, (Center for Study of Ethical Development about the DIT, 2019; Narvaez & Bock, 2002). Importantly, the DIT effectively measures how individual's schemas change when addressing "macro" questions, such as how to organize society-wide cooperation (Rest et al., 1999).

Three measurement schemas factor into DIT scores: Personal Interest Schema, Maintaining Norms Schema, and Postconventional Schema, formally categorized as Kohlberg Stage 2 and 3, Stage 4, and Stage 5 and 6, respectively (Narvaez & Bock, 2002). These three schemas represent mental models used in reasoning about moral dilemmas (Narvaez & Bock, 2002; Rest et al., 1999). Notably, the DIT measures tacit knowledge and provides post-dilemma stimuli in the forms of fragments of reasoning. It has been recognized as an important instrument in measuring moral judgement (Department of the Army, 2008).

The complete DIT-1 consists of six dilemmas and takes about 45–50 minutes to complete, whereas the DIT-2 consists of five dilemmas and takes about 40–45 minutes. There is a short form of both DITs, with only three stories, taking about 35–40 minutes to complete, although the shortened test generally lowers there liability and power of validity trends (Center for Study of Ethical Development About the DIT, 2019). The traditional short form of the DIT-1 consists of the first three stories; the recommended combination for short form use with the DIT-2 is stories 1, 2, and 4. For the purposes of this research plan, we will use the shortened DIT-2 as it contains more contemporary moral dilemmas, and to reduce length of the experiments.

6.3 Moral Sensitivity: Life Events Instrument

Increasingly, moral sensitivity has garnered attention within ethical decision making. This includes the recognition and perceived importance of an ethical issue, ethical judgment, and ethical intention (MacIntyre, Doty, & Hu, 2016). For example, Valentine and Godkin (2019) demonstrated that there is a positive correlation between an individual's ability to recognize and perceive the importance of an ethical issue and ethical judgment to whistleblowing activity. This work builds on the assertions of Rest et al. (1999) that ethical behavior is a multifaceted phenomenon consisting of four interrelated psychological components, the first of which includes ethical sensitivity. Sirin, Rogers-Sirin, and Collins (2010) note that ethical sensitivity is the most critical building block in formulating ethically defensible courses of action when faced with instances of intolerance. MacIntyre, Doty, and Xu (2016) identify the unlawful behavior of US Military at Abu Ghraib prison during the Iraq war as an example of a lack of ethical thinking and sensitivity.

However, MacIntyre, Doty, and Xu (2016) note that although thinking morally or ethically may not come naturally, it can be developed using as a starting point the development of an individual's awareness and identification that an ethical issue exists. In this effort, they have developed an Ethical Sensitivity Mindfulness Model (ESMM) as a comprehensive approach to understanding ethical sensitivity. This conceptual model focuses on the "Degree of Mindfulness" of a participant. Essentially, a mindless state precludes ethical sensitivity, where more mindful individuals have a greater likelihood of being ethically sensitive. In their words:

> The Cognitive/Affective Processes and Mindfulness boxes in our model are also informed by the Elaboration Likelihood Model of persuasion, which suggests individuals have and use both central and peripheral routes in their thinking. These routes are influenced by the individual's ability to process information, their motivation to process the information, and the strength or clarity of the information. The ability to meta-cognize, challenge one's thinking, be conscious of one's thinking, and correct errors in one's thinking, all speak to the spectrum and variety available in the Mindfulness box (the human brain at work).
>
> (MacIntyre, Doty, & Hu, 2016)

Accordingly, to measure the ethical sensitivity of an individual, MacIntyre, Doty, and Xu (2016) created the *Life Events* instrument. The value of this instrument lies in part to the fact that it disguises its central interest in the participant's ethical sensitivity. The limitation of this instrument is that it has not yet been validated (preliminary work is underway). Overall, we believe integrating a measure of ethical sensitivity into the design and lexicon of a virtual human agent holds great promise to further refine how a virtual human system can detect and respond to one's ethical sensitivity, triangulating those data with personality measures and an assessment of moral development.

7 Conclusion

In the face of an increasing presence of human virtual agents in a range of contexts, particularly in the domain of education and training in the military, there is a compelling argument to make a more concerted effort to design robust virtual human agents aimed at supporting the development of ethical decision making for military personnel. Of particular interest is leveraging insights found in the area of personality research, moral judgment, and moral sensitivity as frameworks to shape emerging, AI-driven human virtual agents. The US Army will increasingly develop and employ AI-driven systems and synthetic training environments to prepare the future warfighter. Notably, the efforts to develop human virtual agents that will serve as monitors and companions to soldiers continues to gain traction within the ranks. In the near future, it is more likely than not that all soldiers on the ground and within the leadership hierarchy will have at their disposal a JARVIS of sorts—an AI-driven human virtual agent who can advise, recommend, and assist in mission tasks. Within that context, we see this integration of ensuring that these systems also support ethical decision making capabilities of end users as an area of immediate and significant need. Of course, this represents an opportunity to expand the use and efficacy of human virtual agents. More significantly, intentional and ethically-driven design of these systems can create an environment in which virtual humans bolster and even augment our organic humanity.

References

Arad, A., & Feige, K. (2008). *Iron man*. Marvel Studios.
Artstein, R., Gandhe, S., Gerten, J., Leuski, A., & Traum, D. (2009). Semi-formal evaluation of conversational characters. In O. Grumberg, M. Kaminski, S. Katz, & S. Wintner (eds.), *Lecture notes in computer science* (Vol. 5533 LNCS, pp. 22–35). Springer. https://doi.org/10.1007/978-3-642-01748-3_2.
Ashton, M. C., & Lee, K. (2007). Empirical, theoretical, and practical advantages of the HEXACO model of personality structure. *Personality and Social Psychology Review*, *11*(2), 150–166.
Ashton, M. C., & Lee, K. (2009). The HEXACO–60: A short measure of the major dimensions of personality. *Journal of Personality Assessment*, *91*(4), 340–345.
Ashton, M. C., Lee, K., Perugini, M., Szarota, P., de Vries, R. E., Di Blas, L., et al. (2004). A six-factor structure of personality-descriptive adjectives: Solutions from psycho-lexical studies in seven languages. *Journal of Personality and Social Psychology*, *86*, 356–366.
Batey, M., & Furnham, A. (2006). Creativity, intelligence, and personality: A critical review of the scattered literature. *Genetic, Social, and General Psychology Monographs*, *132*(4), 355–429.
Baur, T., Damian, I., Gebhard, P., Porayska-Pomsta, K., & Andre, E. (2013). A job interview simulation: Social cue-based interaction with a virtual character. In *2013 ASE/IEEE International Conference on Social Computing*, pp. 220–227.
Baylor, A. L. (2011). The design of motivational agents and avatars. *Educational Technology Research and Development*, *59*(2), 291–300. https://doi.org/10.1007/s11423-011-9196-3.
Brémault-Phillips, S., Pike, A., Scarcella, F., & Cherwick, T. (2019). Spirituality and moral injury among military personnel: A mini-review. *Frontiers in Psychiatry*, *10*, 276.
Campbell, J., Hays, M. J., Core, M., Birch, M., Bosack, M., & Clark, R. E. (2011). Interpersonal and leadership skills: Using virtual humans to teach new officers. In *Interservice/Industry Training, Simulation, and Education Conference (I/ITSEC) 2011*. Orlando, FL.
Center for Study of Ethical Development About the DIT. (2019). https://ethicaldevelopment.ua.edu/about-the-dit.html.
Chamorro-Premuzic, T., & Furnham, A. (2008). Personality, intelligence and approaches to learning as predictors of academic performance. *Personality and Individual Differences*, *44*(7), 1596–1603.

Chattaraman, V., Kwon, W.-S. S., & Gilbert, J. E. (2012). Virtual agents in retail web sites: Benefits of simulated social interaction for older users. *Computers in Human Behavior, 28*(6), 2055–2066. https://doi.org/10.1016/j.chb.2012.06.009.

Chollet, M., Morency, L., Shapiro, A., Scherer, S., & Angeles, L. (2015). Exploring feedback strategies to improve public speaking: An Interactive virtual audience framework. In *Ubi Comp '15* (pp. 1143–1154). ACM. https://doi.org/10.1145/2750858.2806060.

Clarke, A. (1975). *Tales from the White Hart.* Ballantine Books.

Currier, J. M., Holland, J. M., & Malott, J. (2015). Moral injury, meaning making, and mental health in returning veterans. *Journal of Clinical Psychology, 71*(3), 229–240.

de Graaff, M. C., Giebels, E., Meijer, D. J., & Verweij, D. E. (2019). Sensemaking in military critical incidents: The impact of moral intensity. *Business & Society, 58*(4), 749–778.

DeFalco, J. & Sinatra, A. (2019). Adaptive instructional systems: The Evolution of hybrid cognitive tools and tutoring systems. In *Proceedings from AIS I: HCII 2019*, pp. 52–61, Orlando, FL.

Department of Army (2008). The U.S. Army study of the human dimension in the future 2015–2024. *TRADOC Pam 525-3-7-01*. Fort Monroe, VA.

Emonet, S. M. F. (2020). The importance of ethics education in military training. https://www.armyupress.army.mil/Portals/7/nco-journal/docs/Emone-Ethics-v4.pdf.

Erdheim, J., Wang, M., & Zickar, M. J. (2006). Linking the big five personality constructs to organizational commitment. *Personality and Individual Differences, 41*(5), 959–970.

Eysenck, H. J., & Eysenck, S. B. G. (1975). *Manual of the Eysenck personality questionnaire.* Hodder & Stoughton.

FM 3-07 Stability. (2014). Washington, DC: Headquarters, Department of Army.

Gratch, J., Rickel, J., André, E., Cassell, J., Petajan, E., & Badler, N. (2002). Creating interactive virtual humans: Some assembly required. *IEEE Intelligent systems, 17*(4), 54–63.

Hart, J. L., & Proctor, M. D. (2016). Framework and assessment of conversational virtual humans as role-players in simulated social encounters with people. *Journal of the International Association of Advanced Technology and Science, 3*(2), 24–33.

Hart, J. L., & Proctor, M. D. (2019). Behaving socially with a virtual human role-player in a simulated counseling session. *The Journal of Defense Modeling and Simulation.* https://doi.org/https://doi.org/10.1177/1548512918825349.

Hatfield, J., Steele, J. P., Riley, R., Keller-Glaze, H., & Fallesen, J. J. (2011). *2010 Center for army leadership annual survey of army leadership (CASAL): Army education (Technical Report 2011–2).* Fort Leavenworth, KS.

Hill Jr, Randall W., James Belanich, H. Chad Lane, Mark Core, Melissa Dixon, Eric Forbell, Julia Kim, and John Hart. (2006). *Pedagogically structured game-based training: Development of the ELECT BiLAT simulation.*

University Of Southern California Marina Del Rey Ca Inst For Creative Technologies.

Johnsen, K. J. (2008). *Design and validation of a virtual human system for interpersonal skills education. Methodology.* University of Florida. https://doi.org/UMI No. 3334476.

Johnsen, K., Raij, A., Stevens, A., Lind, D. S., & Lok, B. (2007). The validity of a virtual human experience for interpersonal skills education. In *Proceedings of the SIGCHI Conference on Human Factors in Computing Systems* (pp. 1049–1058).

Johnson, W. L. (2015). Cultural training as behavior change. In *6th International Conference on Applied Human Factors and Ergonomics (AHFE 2015)* (pp. 1–8).

Johnson, W. L., & Valente, A. (2008). Tactical language and culture training systems: Using artificial intelligence to teach foreign languages and cultures. In *AAAI* (pp. 1632–1639).

Jones, S. G., & Muñoz, A. (2010). *Afghanistan's local war: Building local defense forces.* RAND Corporation.

Kavathatzopoulos, I. (1993). Development of a cognitive skill in solving business ethics problems: The effect of instruction. *Journal of Business Ethics, 12*(5), 379–386.

Kenny, P., Hartholt, A., Gratch, J., Swartout, W., Traum, D., Marsella, S., & Diane, P. (2007). Building interactive virtual humans for training environments. In *Interservice/Industry Training, Simulation, and Education Conference (I/ITSEC).* Orlando, FL.

Kilcullen, D. (2009). *The accidental guerrilla.* Oxford University Press.

Kramer, E. H., van Bezooijen, B., & Delahaij, R. (2010). Sensemaking during operations and incidents. *Managing Military Organizations: Theory and Practice,* 126.

Landgren, J. (2005). Supporting fire crew sensemaking enroute to incidents. *International Journal of Emergency Management, 2,* 176–188.

MacIntyre, A., Doty, J., & Xu, D. (2016). Ethical sensitivity during military operations: Without mindfulness there is no sensitivity. In S. Belanger & D. Lagace-Roy (eds.), *Military ethics and well-being.* McGill-Queen's University Press., pp. 143–171.

Marcus, J., & Roy, J. (2019). In search of sustainable behaviour: The role of core values and personality traits. *Journal of Business Ethics, 158*(1), 63–79.

Markoff, J. (2011). Computer wins on "Jeopardy!": Trivial, it's not. *NYTimes.Com.*

Martindale, C., & Dailey, A. (1996). Creativity, primary process cognition and personality. *Personality and Individual Differences, 20*(4), 409–414.

Micewski, E. R. (2016). Conveying ideas and values in education! Challenges in teaching military ethics. In T. R. Elssner, R. Janke, & A. C. Oesterle (eds.), *Didactics of military ethics* (pp. 173–177). Brill Nijhoff.

Molendijk, T. (2019). The role of political practices in moral injury: A study of Afghanistan veterans. *Political Psychology, 40*(2), 261–275.

Narvaez, D., & Bock, T. (2002). Moral schemas and tacit judgement or how the defining issues test is supported by cognitive science. *Journal of Moral Education, 31*(3), 297–314.

Park, M., Kjervik, D., Crandell, J., & Oermann, M. H. (2012). The relationship of ethics education to moral sensitivity and moral reasoning skills of nursing students. *Nursing Ethics, 19*(4), 568–580.

Reed, G. S., Petty, M. D., Jones, N. J., Morris, A. W., Ballenger, J. P., & Delugach, H. S. (2016). A principles-based model of ethical considerations in military decision making. *The Journal of Defense Modeling and Simulation, 13*(2), 195–211.

Rest, J. R., Narvaez, D., Thoma, S. J., & Bebeau, M. J. (1999). DIT2: Devising and testing a revised instrument of moral judgment. *Journal of Educational Psychology, 91*(4), 644.

Rizzo, A., Lucas, G., Gratch, J., Stratou, G., Morency, L.-P., Shilling, R., Scherer, S. (2016). Clinical interviewing by a virtual human agent with automatic behavior analysis. In *Proceedings of 11th International Conference on Disability, Virtual Reality & Associated Technologies* (pp. 57–64). University of Reading.

Robinson, P., De Lee, N., & Carrick, D. (2008). *Ethics education in the military*. Ashgate Publishing Company.

Rudman, P., & Zajicek, M. (2006). Autonomous agent as helper—Helpful or annoying? In *2006 IEEE/WIC/ACM International Conference on Intelligent Agent Technology* (pp. 170–176). IEEE Computer Society Washington, DC. https://doi.org/10.1109/IAT.2006.41.

Selart, M., & Johansen, S. T. (2011). Ethical decision making in organizations: The role of leadership stress. *Journal of Business Ethics, 99*(2), 129–143.

Shields, G. S., Sazma, M. A., & Yonelinas, A. P. (2016). The effects of acute stress on core executive functions: A meta-analysis and comparison with cortisol. *Neuroscience & Biobehavioral Reviews, 68*, 651–668.

Silvia, P. J., Martin, C., & Nusbaum, E. C. (2009). A snapshot of creativity: Evaluating a quick and simple method for assessing divergent thinking. *Thinking Skills and Creativity, 4*(2), 79–85.

Simkins, D. W., & Steinkuehler, C. (2008). Critical ethical reasoning and role-play. *Games and Culture, 3*(3–4), 333–355.

Sirin, S., Rogers-Sirin, L., & Collins, B. (2010). A measure of cultural competence as an ethical responsibility: Quick-racial and ethical sensitivity test. *Journal of Moral Education, 1*(39), 49–64.

Swartout, W. (2010). Learned from virtual humans. *AI Magazine, Spring 2010*, 9–20.

Swartout, W., Traum, D., Artstein, R., Noren, D., Debevec, P., Bronnenkant, K., White, K. (2010). Ada and Grace: Toward realistic and engaging virtual museum guides. In *Intelligent Virtual Agents 2010* (pp. 286–300), Springer.

Tartaro, A., Cassell, J., Ratz, C., Lira, J., & Nanclares-Nogués, V. (2014). Accessing peer social interaction: Using authorable virtual peer technology as a component of a group social skills intervention program. *ACM Transactions on Accessible Computing, 6*(1), 2–29. https://doi.org/10.1145/2700434.

Thoma, S. J., & Dong, Y. (2014). The Defining Issues Test of moral judgment development. *Behavioral Development Bulletin, 19*(3), 55.

Traum, D. (2008). Talking to virtual humans: Dialogue models and methodologies for embodied conversational agents. In I. Wachsmuth & G. Knoblich (eds.), *Modeling communication with robots and virtual humans* (pp. 296–309). Springer-Verlag.

Traum, D., Roque, A., Leuski, A., Georgiou, P., Gerten, J., Martinovski, B., Vaswani, A. (2007). Hassan: A virtual human for tactical questioning. In *Proceedings of the 8th SIGdial Workshop on Discourse and Dialogue* (pp. 71–74).

Turing, A. M. (1950). Computing machinery and intelligence. *Mind*, 59(236), 433–460. https://doi.org/10.1093/mind/LIX.236.529-b.

Tybur, J. M., & de Vries, R. E. (2013). Disgust sensitivity and the HEXACO model of personality. *Personality and Individual Differences*, 55(6), 660–665.

Valentine, S., & Godkin, L. (2019). Moral intensity, ethical decision making, and whistleblowing intention. *Journal of Business Research*, 98, 277–288.

Von Der Pütten, A. M., Krämer, N. C., Gratch, J., & Kang, S. H. (2010). "It doesn't matter what you are!" Explaining social effects of agents and avatars. *Computers in Human Behavior*, 26(6), 1641–1650. https://doi.org/10.1016/j.chb.2010.06.012.

Wansbury, T. G., Hill, R. W., & Belman, O. (2014). ELITE training. *Army AL&T Magazine*, (October–December), 56–61.

Warner, C. H., Appenzeller, G. N., Mobbs, A., Parker, J. R., Warner, C. M., Grieger, T., & Hoge, C. W. (2011). Effectiveness of battlefield-ethics training during combat deployment: A programme assessment. *The Lancet*, 378(9794), 915–924.

Weick, K. E., Sutcliffe, K. M., & Obstfeld, D. (2005). Organizing and the process of sensemaking. *Organization Science*, 16(4), 409–421.

Weizenbaum, J. (1966). ELIZA–A Computer program for the study of natural language communication between man and machine. *Communications of the ACM*, 9(1), 36–45.

Wilson, K. E., Lowe, M. X., Ruppel, J., Pratt, J., & Ferber, S. (2016). The scope of no return: Openness predicts the spatial distribution of inhibition of return. *Attention, Perception, & Psychophysics*, 78(1), 209–217.

Yates, W. (2011). Tactical Iraqi language and cultural training systems: Lessons learned from 3rd battalion 7th Marines 2007. In *Presentation at Defense GameTech*. Orlando, FL.

Chapter 8

Ethical Frameworks for Cybersecurity

Applications for Human and Artificial Agents

F. Jordan Richard Schoenherr and Robert Thomson

1 Introduction

Humans are an adaptive species. As we move from physical to virtual information domains, we face both familiar and novel social and ethical challenges (see Schoenherr, 2022). With increasing concerns of adversarial cyberoperations, defensive cybersecurity, and multinational cyberwar, there is a growing need to adapt old strategies to new environments. Despite innovations in hardware and software that define these domains, cybersecurity is fundamentally based on social and cognitive processes (e.g., Beskow & Carley, 2019). Perhaps the most important of these are norms and conventions associated with interpersonal interaction related to ethical and social issues. Recently, professional organizations associated with software development have emphasized the need to develop and integrate ethical norms into autonomous and intelligent systems (A/IS)[1] in order to ensure that these systems serve humans (e.g., IEEE and ACM; Schoenherr & Burleigh, 2020). For the purposes of this chapter, we consider machine learning and artificial intelligence to fall under the A/IS umbrella. In that cyberwarfare techniques are associated with the use of A/IS (e.g., malware, social media bots), serious consideration needs to be given to how best to adapt theories from the behavioral and social sciences to this new ethical domain.

Discussion of cyber ethics (e.g., Spinello & Tavani, 2005) and cyberwarfare in particular (Allhoff et al., 2016) tend to focus on specific issues (e.g., anonymity, privacy, intellectual property, and regulation) but often fail to adequately address the social cognitive processes that are associated with identifying and responding to these issues (cf. Schoenherr, forthcoming; Schoenherr & Thomson, 2020). Moreover, social cognitive processes are both informed by, and modify, mental representations of fairness in terms of social scripts, schemata, and relational models (i.e., Relational Models Theory; Fiske, 1992; Rai & Fiske, 2011). Collectively, these mental representations provide schematic frameworks

DOI: 10.4324/9781003030928-12

(or, frames) for understanding the ethical affordance of an environment. Here, we consider ethical affordance in terms of the ethically relevant features and associated responses that are available to an individual within a social environment. These frames set the expectations for users, thereby determining their behavior such as cooperation and competition, rewards and punishment (e.g., Fiske & Rai, 2014). If these frames are as universally applicable as researchers in the behavioral and social sciences claim, they can provide insight into users' behavior in virtual domains that can help inform the development of cybersecurity strategies and A/IS development more generally.

In order to consider the usefulness of ethical frames in providing insight into cybersecurity, this chapter reviews the cognitive and social cognitive frames that can inform cybersecurity. In the first section, we consider the widespread use of analogies used to understand cybersecurity, e.g., "cyberwar", "cyber Hiroshima" (de Matos Alves, 2015; Sulek et al., 2009). Despite the utility of analogical reasoning and its ubiquity in the information and computer sciences (e.g., "desktop", "mouse", "web", "virus"), analogies can both facilitate and interfere with our understanding of a novel situation (Neustadt & May, 1986). In the second section, we consider one pervasive frame (security as a game) and provide a more exhaustive schema-based framework that can incorporate traditional game theoretic approaches (e.g., the Prisoner's Dilemma) along with a variety of other social dilemmas and exchange norms. This approach describes abstract relational structures (e.g., equality or dominance), the appropriate exchange norms, and their pay-offs given patterns of cooperation and competition. In that these models can be implemented computationally, we demonstrate how they can be used to inform network security strategies and assist in the development of A/IS.

2 Analogies of Information Domains

The use of analogies in cybersecurity and cyberoperations is not surprising given the importance of analogical reasoning in learning and decision-making (Brown & Clement, 1989; Gick & Holyoak, 1980; Rumelhart & Abrahamson, 1973; Sternberg, 1977). For instance, reasoning by analogy is widely used in the natural and social sciences (e.g., Dunbar, 1997; Hoffman, 1980; for a review in the context of A/IS, see Schoenherr & Thomson, 2020). Analogical reasoning provides a means to highlight and understand structural relationships between a known and an unknown domain in terms of similarity (Vosniadou & Ortony, 1989). Although experimental settings have often suggested strict limitations on the "distance" between domains (i.e., the dissimilarity between a novel and familiar domain; Dunbar, 2001), ecological studies demonstrate that complex analogies can be used to understand novel domains (Blanchette & Dunbar, 1997). In the context of ambiguous information domains, analogies can assist in facilitating comprehension (Burstein, 1983; Clement et al., 1986; Pea & Kurland, 1984).

The most widespread analogy in the information and computer sciences is the notion of cyberspace. From its origins in science fiction, cyberspace represented an analogy based on an understanding of physical space (Gibson, 1984):

> [Cyberspace is a] consensual hallucination by billions... a graphic representation of data abstracted from the banks of every computer in the human system. Unthinkable complexity. Lines of light ranged in the nonspace of the mind, clusters and constellations of data. Like city lights, receding.
>
> (p. 51)

Gibson's description of cyberspace provides the reader with a more concrete representation of an abstract idea, allowing them to better understand and predict what might happen within such settings. Crucially, analogies are effective in facilitating performance to the extent that they use a known domain to emphasize and deemphasize similarities within a target domain (Chen et al., 2004). However, as Neustadt and May (1986) have suggested, the adoption of analogies can interfere with decision-making when similarities and dissimilarities between the actual and target domains are not considered, leading to oversimplification and misinterpretation. As in the case of cyberspace, the analogies used to understand interactions within cyberspace need to be critical considered in order to identify the trade-offs inherent in their adoption.

2.1 Cyberoperations and Cyberwar

In the same manner as cyberspace, analogies used to understand cyberoperations are associated with the same issues (Sulek et al., 2009). One of the most frequently used metaphors is "cyberwar" introduced by Arquilla and Ronfeldt (1993). "Cyberwar" reflects a military-level conflict between state actors. The concept of cyberwar has also spawned several related concepts. For instance, in an analogous manner to land power, sea power, and air power, "cyberpower" has also been the focus of considerable discussion (e.g., Betz, 2017; Jordan, 1999; Nye, 2010). Cyberpower can be defined as the ability to operate effectively in cyberspace, including protecting one's own online assets and engaging adversaries in cyberspace. The very nature of cyberspace is that it does not necessarily have a local physical presence, and the notion of cyberpower assumes a quantifiable metric in a way that manpower or technical superiority can be quantified.

In a review, Rattray (2009) considers environmental power, paying particular attention to air and space power. He identifies four analogical similarity dimensions: technological advances, speed and scope of operations, control of key features, and national mobilization. In terms of technological advances, he emphasizes a thread of technological evolution:

> The development of airpower in the first half of the 20th century meant that attacks could be launched against strategic centers of gravity in hours. The advent of ballistic missiles with nuclear warheads after World War II brought timelines down to minutes, and the scale of effects rose dramatically. The cyberspace of the 21st century means key events and disruptive threats can necessitate responses in seconds.
>
> (p.267)

He additionally uses the concept of spacepower to highlight other affordances that are relevant to cyberpower, i.e., "[c]ontinuous operations with no territorial limits would be an inherent feature of the space environment, creating a new high ground that would enable those with enough spacepower to dominate operations in other environments" (Rattray, 2009; p. 266).

Similarly, numerous specific analogies and analogical frameworks have been provided to understand cyberoperations, each highlight different affordances (e.g., Goldman & Arquilla, 2014; Perkovich & Levite, 2017; see Table 8.1). For instance, Arquilla and Ronfeldt (1993) suggest that "...cyberwar may be to the 21st century what blitzkrieg was to the 20th century" (p. 31). Betz and Stevens (2013) consider spatial and biological analogies that have been used to understand cybersecurity. Other broad analogies have used fear-inducing events from history including "Cyber-Katrina" (Lauland, 2016), "Digital 9/11" (Kroll, 2019), "cyber Chernobyl" (Myers, 2011), and "cyber Armageddon" (Lucky, 2010; for reviews, see de Matos Alves, 2015; Sulek et al., 2009. More neutral conflict-based analogies have also been used such as describing cyberoperations as a "force multiplier" (Thomas, 2006) likening them to analogs in kinetic operations or guerilla warfare, which are much smaller in scale and restricted in scope (Christensen, 1999; Liles, 2010; Liles & Rogers, 2008; van Haasster et al., 2016). Alternatively, social and societal changes that occurred within the nuclear era have also been used to highlight the dual-uses of this technology (Nye, 2011). Many other analogies have also been proposed, often with minimal detail or consideration (e.g., Axelrod, 2014; Perkovich & Levite, 2017).

An overriding issue related to the adoption of cyberwar as an analogy is that cyberoperations are not conducted against adversaries with which we are at war with in a tradition sense. Recently, scholars have highlighted the importance of this distinction given that the ethics of warfare are applicable to cyber operations (e.g., Arquilla, 2013; Nguyen, 2013). Partially addressing this point, others refer to operations as a "cyber cold war" to highlight the adversarial nature of the interactions while falling short of full war (Sjouwerman, 2020). To this end, analogies related to "war" might reflect disanalogies that suggest inappropriate responses.

2.2 Kriegsspiel: The Game Analogy and Game Theory

In that analogies and metaphors help us understand a system and aid in its prediction, their specificity must also be considered. For instance, while "cyberwar" provides a general idea about the nature of conflict between two large social organizations (i.e., states), it neither suggests what kind of warfare will occur nor the strategies and tactics that would be used. While "cold war", "blitzkrieg", and "Chevauchées" are more specific, they do not provide sufficient insight that allows for the kinds of predictions required to confront malicious autonomous and intelligent systems (A/IS). Sproull and Kiesler (1991) did not create the term cyberwar to provide specific guidance for state and nonstate actors. They note that "[t]he full range of *payoffs*, and the *dilemmas*, will come from how the technologies affect how people can think and work together" (italics added, p. 16).

Games have been a ubiquitous feature of military strategy (e.g., Prussian army's use of *kriegsspiel*, or war games). The most prominent adaptation to this approach is evidenced in Game Theory, a mathematical approach to help understand interpersonal interactions (Myerson, 1997). In contrast to other analogies, Game Theory provides a more specific *prescriptive* model of human behavior that assumes that social situations can be understood in an analogous manner to games with players, rules, and points (von Neumann & Morgenstein, 1947).[2] Game theory assumes that players *should* make rational decisions that maximize their outcomes in these games given a specific set of rules. Nash (1950) provides one prominent means to understand the individual rationality that informs the decision-making process, what he defined as a game's equilibria. A Nash equilibrium reflects a state in the game when player's decisions should remain unchanged due to the optimization of their respective outcomes, i.e., a player cannot attain any greater outcome by making an alternative choice.

In general, game theoretic models applied to security assumes competition between actors (Tambe, 2012), the simplest of which is a two-person (Player A and Player B), two-option (cooperate or defect) game. For instance, Stackelberg Competitions (von Stackelberg, 1934; Yin et al., 2010) have often been used to conceptualize, and respond to, security issues. In economics, these competitions assume players are defined by one of two roles: leaders and followers. In the context of security, players instead take on roles of defender and adversary. The game unfolds in phases with the defender (leader) and attacker (follower) taking turns allocating their resources. Given the limited resources of the attacker and the defender, they must decide how to distribute themselves to optimally defend or attack their opponent. In cybersecurity, this might involve a defender establishing certain network defence operations using honeypots or other optimal resource allocation for selecting which network traffic packets to evaluate on a given network (Kar et al., 2018; Sedjelmaci, et al., 2018; Zhu & Martínez, 2011).

Whether players *do* or *should* compete with one another remains a problematic assumption. In the classic demonstration of this, the Prisoner's Dilemma, the collective interest of two players (i.e., cooperate) is placed in competition with the self-interest of each individual player (i.e., defect). In practice, this is implemented by providing a pay-off matrix (see Figure 8.1). The Prisoner's Dilemma is one of many social dilemmas defined by pay-off matrices (Balliet et al., 2017; Kelley &Thibaut, 1978; for examples, see Table 8.1). Again, the assumption is that players should make choices that maximize their outcomes. For instance, the Nash equilibrium in the Prisoner's Dilemma is where both players choose to defect: Player 1 receives the greatest reward if they defect, as would Player 2 (Nash, 1950). If the assumptions of Game Theory hold, they provide a useful resource for A/IS designers as they can be readily implemented computationally in A/IS that operate in adversarial environments or network defence.

The assumptions concerning mutual distrust and antagonism reflected in the statements of many state and non-state actors would appear to

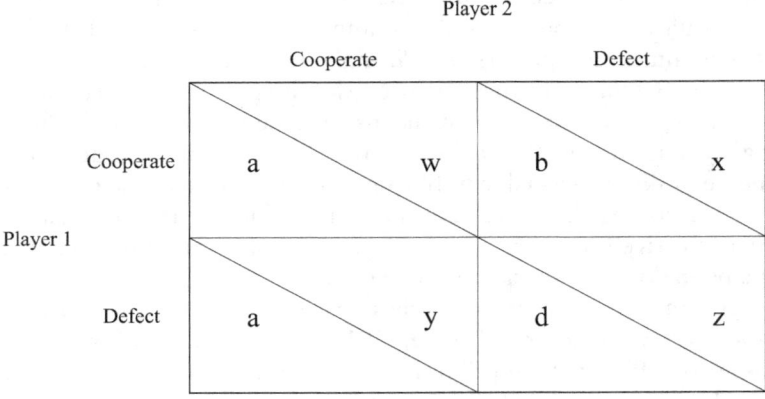

Figure 8.1 General structure of a two-person, two-option social dilemmas.

Table 8.1 Analogies for cyberoperations

Analogy Type	Analogy	Similarity Features and Dimension(s)	Exemplars
Military/ Aggression	Air power/ Defense	Speed. Technological advancement. Lessening of restrictions. Conferred advantage.	Airplanes. Missile systems.
	Arms race	Weapons build-up. Reactionary. Costly. Repurposing of technology. Redirecting of labor.	Control of chemical weapons. Nuclear arms race.
	Concerted strike	Balance of offensive/defensive power. Physical disruption/incapacitation. Reversal. Coordinated by a state actor.	Battle of Britain. Chevauchées.
	Cold War	Multiple fronts. Indirect attacks. Little to no attribution.	Cuban missile crisis. Korean War.
	Surprise attack.	Fast. Surprise. Coordinated. Aggressive. Debilitating.	Blitzkrieg. Pearl Harbor. Tet Offensive.
	Force Multiplier	Coordinated by state actors. Financial disruption/incapacitation. Nonlethal, non-military. Increases/facilitates effect of primary operations.	Artillery. Economic Warfare.

(Continued)

Analogy Type	Analogy	Similarity Features and Dimension(s)	Exemplars
	Guerilla/ Submarine Warfare	Small strikes. Surprise. Atypical method of attack. Imbalance in power. Different strategies of engagement. Coordinated by non-state actors.	Trojan Horse. 9/11.
	Nonlethal weapons	Weapon. No death results from their use. Temporary disruption/incapacity.	Rubber bullets. Stun gun.
	Precision-Guided Munitions	Autonomous. Attempted localized effect. Imprecise despite expectations.	Drones. Missiles.
	Nuclear Weapons/ Nuclear Arms	Destructive. Surprise. Threats. Devastating impact. Debilitating. Brinksmanship. Arms race.	Mutually Assured Destruction (MAD). Hiroshima.
	Privateering/ Piracy	Aggressive. Theft. Small scale strikes. Illegal.	Boat-borne attacks. Cyber criminals.
	Subterfuge	Denial of attack. Denial of intent. Strategic. Black/grey hat.	Manchurian railway (WWII). False flag operations.

support the use of Prisoner's Dilemma in predicting behavior in information domains. However, even the earliest empirical results do not support the claim that players will uniformly defect in order to maximize their own gains (Rapoport & Chammah, 1965). Numerous contemporary studies have demonstrated that the underlying assumptions of mutual defection suggested by Nash and others do not hold. Such results are typically explained in terms of failures of human rationality as the result of error. However, the alternative possibility is that the adversarial assumptions do not accurately describe a typical player's assumptions when approaching an exchange situation (Balliet et al., 2017; Kelley & Thibaut, 1978). This suggests that other systematic biases influence players' responses or that players' assumptions when engaging in such tasks are wholly different.

Systematic differences between behavior predicted by theories of rational choice and observed in participant performance in these tasks have been referred to as a description-experience gap (Wulff et al., 2018). The description-experience gap assumes that there can be a conflict between the stated structure of the game (i.e., the description) and a player's own beliefs based on past interactions with other players, i.e., their experience. In addition to external influences, players can also update their expectations throughout the course of the playing (Lee, 1971; Luce & Suppes, 1965). For instance, the availability of reputational information (i.e., whether someone cooperated on a previous round of an iterative game) will affect interactions (Martin et al., 2014), if the task is framed in terms of "community" interaction rather than "business" interaction will increase cooperation (Liberman et al. 2004; Pillutla & Chen, 1999; Tenbrunsel & Messick, 1999), how rewards will be allocated (Grinberg et al. 2012), and the strength of the relationship maintained between players (Harrison et al., 2011; Xu et al., 2011). These findings suggest that multiple relationship schemata are available. Consequently, if models of cybersecurity solely focus on adversarial relationships, they will not provide adequate predictions of typical users. Instead, other models of social situations should be sought.

2.3 Schematic Models of Relational Structures

An alternative approach is to extend the number of possible structures of social situations and account for the corresponding social forces that motivate players. For Lewin (1939; for a recent discussion, see Burnes & Cooke, 2013), humans live within a field of social forces that are analogous to physical forces. Along these lines, Interdependence Theory (Balliet et al., 2017; Kelley & Thibaut, 1978) assumes that a number of generalizable social relationships exist and that each can be defined in terms of a pay-off matrix. Consequently, if individuals fail to conform to the rational behavior prescribed by the Prisoner's Dilemma, the observed description-experience gap might be attributable to a participant's belief that the situation is more accurately represented by another social dilemma defined by a different pay-off matrix.

Interdependence Theory provides a useful resource for understanding the variety of interpersonal situations that we can encounter in our exchanges with others. In addition to assuming that reward and punishment will determine behavior, it assumes that a pay-off matrix can be *decomposed* into separable components that control the outcomes of a situation. Namely, the decisions made by a player (*Actor Control*), their partner (*Partner Control*), and their conjoint decisions (*Partner Control*) will all determine the outcomes for each player. Interdependence Theory assumes that Actor Control, Partner Control, and Joint Control provide a means to understand the forces that affect an agent's decisions. For instance, Schoenherr and Thomson (2020) have noted that this computational framework can be adapted to network security and that these variance components can be tested to model human behavior.

In order to understand how pay-off matrices change motivation, consider two common social dilemmas (see Table 8.2): the Assurance Dilemma and the Chicken Dilemma (Halevy et al., 2012; Messick, 1999; Rapoport & Guyer, 1966; Schoenherr & Thomson, 2020). In the Assurance Dilemma, individual and collective rewards are maximized when both parties cooperate. For instance, two nation-states benefit collectively and individually if they both develop network security technologies. In the Chicken Dilemma, individual reward is greatest when another player defects, while collective reward is greatest when both players defect. For instance, if two nation-states engage in adversarial cyberoperations, cessation of operations by one state would be beneficial to another and detrimental to themselves.

An important observation in subsequent extensions in Interdependence Theory is that players can *transform* the pay-off matrix that is provided by an experimenter into one that they believe is more suitable. Corresponding to a similar logic of description-experience gap, while a player might be presented with one payoff matrix (e.g., the Prisoner's Dilemma) they might instead behave in a manner that reflects another payoff matrix (e.g., the Assurance Dilemma). Thus, when a description-experience gap is observed, Interdependence Theory suggests that players' responses can be used to infer what social dilemma players believe they are being presented with.

Table 8.2 Payoff structure and types of social dilemmas

Type of Social Dilemma	Player 1	Player 2
Prisoner's Dilemma	$c > a > d > b$	$x > w > z > y$
Assurance Dilemma	$a > c = d > b$	$w > x \geq z > y$
Chicken Dilemma	$b > d > c > a$	$y > z > x > w$

Finally, caution must be taken when adapting Interdependence Theory in any domain. For instance, like the Covariation Model of Attribution also developed by Kelley (1967, 1973, 1980) claims that humans consider the variance components of a situation to determine whether an internal (agent-based) or external (situation-based) attributions are appropriate, it is also unlikely that humans make assessment of control and dependency in this manner. Even early studies demonstrated that attributions are not primarily dependent on these processes (Fletcher, 1983; Lewis, 1995; Passer et al., 1978). Similarly, humans are not naïve scientists that weigh evidence as the model implies. Consequently, although pay-off matrices can be decomposed into basic variance components (Actor Control, Partner Control, and Joint Control) which can be used to derive more complex variance components (see Schoenherr & Thomson, 2020), these claims require more empirical evidence. Moreover, when adapting Interdependence Theory into A/IS and domains to understand human behavior, we must also account for the fact that online interactions are associated with fewer social cues (e.g., voice inflection, body language) that might result in the experience of anonymity. Given that anonymity can reduce prosocial behavior and increase antisocial behavior (e.g., Postmes & Spears., 1998; see, Schoenherr, 2022), the applicability of Interdependence Theory needs to be formally assessed. Nevertheless, this comprehensive theory sets out numerous empirical predictions that can be tested and modeled.

By adopting a schema-based framework for social interaction, developers now have a straightforward means to computationally model social situations. Namely, in contrast to analogy-based frameworks that are inherently idiosyncratic translations between two domains, approaches such as Interdependence Theory provide a systematic framework that can model numerous social situations. Moreover, due to generational and cross-cultural differences in knowledge, it is not necessarily the case that the behavior of network users will be informed by analogies proposed by cybersecurity professionals, e.g., "Millennials" might not be familiar with, or agree, that cyberoperations are analogous to those in the Cold War, Blitzkrieg, or Chevauchées. Consequently, such approaches can be implemented in a straightforward manner when developing ethical A/IS or A/IS that can identify and respond to ethical affordances of a situation (see, Schoenherr, forthcoming).

2.4 Applying Interactional Schemata to Social Agent Development and Detection

A crucial benefit of shifting away from ill-defined analogies to a schema-based framework such as Interdependence Theory is that it can be implemented in a computational framework in a straightforward manner, thereby directly informing the design of A/IS. Even if humans do not adhere to the interactional patterns defined in Interdependence Theory, it will take A/IS closer to a repertoire of human-like exchanges. In order to illustrate this, we briefly consider two applications of relational schemata: beneficent virtual agents such as chatbots, help bots, and programs for detecting malicious bots that operate within networks, e.g., social media chatbots (e.g., Briscoe et al., 2014; Wagner et al., 2012; Wald et al., 2013).

Trust and Exchange with Virtual Agents. Reciprocal interactions depend on, and promote, trust in social agents. Exchange schemata such as Interdependence Theory (Kelley &Thibaut, 1978) are also based on trust (e.g., mutual cooperation in the Prisoner's Dilemma), thus satisfying the expectations of a relational model. Within a social network, trust requires that relevant and accurate information is presented, that individuals adhere to social norms and conventions, and that they have represented themselves in an authentic manner. In that A/IS reflect artificial agents, they must also be accurate, comply with norms and conventions, and present themselves in an authentic manner.

The requirement of authenticity has additional importance given how *human realism* can affect trust in A/IS. Studies suggest that artificial systems that represent imperfect or incomplete depictions of humans can lead to negative affective responses, likely attributable to different levels of familiarity (e.g., Burleigh et al., 2013; Cheetham et al., 2011; Schoenherr & Burleigh, 2020; Schoenherr & Thomson, 2020). Moreover, studies also suggest that the realism of virtual agent behavior is important for trust, cooperation, and sustained interaction (Berkeley et al., 2015; Kiesler et al., 1996; Merritt, & McGee, 2012; Oudah et al., 2015). For instance, using an iterative Prisoner's Dilemma game, Ishowo-Olko et al. (2019) paired users with human or nonhuman players and found that when the nonhuman player were believed to be human, human players were more likely to cooperate then when they believed that the other player was a nonhuman agent. Thus, failures to produce human-like exchange behavior have the potential to reduce cooperation and trust in a user. Consequently, when developing virtual agents that require a user's trust, relational models and social dilemmas can provide a plausible basis for simulating realistic exchange behavior between humans. For instance, in the context of mental health, users of a virtual help agent might expect a simple reciprocal exchange (e.g., peer-peer) or hierarchical relationship (e.g., patient-doctor; e.g., as Relational Models Theory; Fiske, 1992; Rai & Fiske, 2011). Alternatively, dissatisfied customers interacting with a customer service chatbot might instead view the situation as adversarial (e.g., Chicken Dilemma) making the goal of the chatbot to shift the user into another schemata (e.g., Assurance Dilemma).

Social Bots and Anomaly Detection. The notable rise of malicious A/IS in virtual environments has resulted in the creation of numerous detection techniques. Reviews have identified a number of malicious bot detection techniques (Feily et al., 2009; Karim et al., 2014; Zeidanloo et al., 2010). For instance, botnet function can be decomposed into three major components (Zeidanloo & Manaf, 2009): Distribution (how bots are distributed within the network), Sign-On (how bots are connected to a BotMaster program), and Searching (how bots search for vulnerable computers). Two primary methods are used: *Signature-based botnet detection* uses the behavior of known bots to identify network intrusions. *Anomaly-based botnet detection* uses the typical network behavior to identify anomalies in terms of atypical system behavior including high volumes of network traffic, high network latencies, or atypical activity on ports that are seldom used. In an analogous manner, Interdependence Theory can be used to detect anomalies: once user behavior is defined in terms of schemata such as Interdependence Theory (e.g., Prisoner's Dilemma, Assurance Dilemma, and Chicken Dilemma), deviations from these patterns can be used to detect the presence of malicious actors.

The prevalence of bots has also increased in social media platforms created by both humans or A/IS (Chu et al., 2010). These bots can be used to spread disinformation, increase the number of followers for a target set of profiles, or impersonate a user for nefarious purposes (Shao et al., 2016). Consequently, detecting bots on social media has become a research priority (Subrahmanian et al., 2016), leading to a number of detection methods attempting to identify spam, fake reviews, social spam, and link farming (Jiang et al., 2016). For instance, in their development of a bot detection mechanism, Efthimion et al. (2018) used length of user names, reposting rate, temporal patterns such as the hour and frequency of posts, sentiment expression, the amount of followers to friends (too many or too few), and message variability that they claim can correctly classify 97.75% of known bots. However, some of these variables were better predictors than others (e.g., geolocation, languages other than English, and if the account had fewer than 30 followers or more than 1,000 friends).

In that behaviors such as likes, shares, commenting, and messaging reflect traces of interpersonal relationships, the use of Interdependence Theory could be adapted to the identification of anomalous activities on social media. User interaction within social media reflects the strength of their relationships. For instance, Granovetter (1973) noted that "The strength of a tie is a (probably linear) combination of the amount of time, the emotional intensity, and intimacy (mutual confiding), and the reciprocal services which characterize the tie" (p. 136). Measures of the strength of social bonds have included shared friends (Shi et al., 2007), frequency (Gilbert et al., 2008; Granovetter, 1973) and recency of interactions (Lin et al., 1978), link sharing on social media (Gilbert & Karahalios, 2009) as well as communication reciprocity (Ferrara et al., 2016; Friedkin,

1980). In a recent study of Twitter data, Varol et al. (2017) attempted to develop a means to detect bots by using a reciprocal score, defined as the fraction of friends who are also followers. They demonstrated that Twitter accounts attributed to humans were more likely to engage in reciprocal behavior than accounts that were attributed to bots. Thus, an absence of reciprocal interaction suggests that an account is inauthentic.

If bots are not concerned with social interaction, then schema-based models can assist in comparable manner to reciprocity norms. Consequently, if Interdependence Theory reflects an adequate account of how human social agents perceive a number of common exchange scenarios, the motivational components defined by Kelley et al. (2003; i.e., Actor Control, Partner Control, and Joint Control) can be used within time-series data to detect social media bots. For instance, while a user's behavior should be subject to changes in cooperative behavior over time depending on how other users interact with an account, bots will likely display more *invariant* behavior suggesting that they are not using one of the many available schemata. This behavior could be described as *a social* behavior in that it fails to conform to existing schemata (Fiske, 1992). Alternatively, if human users of social media perceive online discussions in terms of these schemata, then exchanges might reflect social dilemmas. For instance, users have the option to agree or disagree with articles or comments and to be civil or uncivil in their replies. This might then have the appearance of a Prisoner's Dilemma, wherein perceived incivility leads to reciprocal incivility. Violations of these norms can then be used to inform detection methods.

3 Conclusion

In that computers provide universal platforms for computing, cyberspace reflects a relatively ambiguous territory. Nevertheless, effective cyberoperations require understanding the affordances of hardware and software as well as the cognitive and social dimensions. Historically, analogical reasoning has helped shape how we perceive these environments, allowing us to make predictions. For instance, the notion of "cyberwar" can help policymakers understand the need to develop attack and defense strategies relative to a given domain, i.e., cyberpower. Caution is required as the use of an inappropriate analogy can have severe negative consequences when used in consequential domains such as national defense (Neustadt & May, 1986). Evocative analogies such as "Cyber Katrina" or "Cyber Hiroshima" might be useful to emphasize the potential suddenness and extent to devastating but ultimately fail to capture the nuances of typical user activity. Adopting more general schema-based frameworks provides a more systematic, theoretically-motivated perspective that can be operationalized into computational models, and can in turn be used to monitor, predict, and defend networks. Using these models, virtual agents can be made to provide more realistic interactions with users and violation of the norms might reflect a useful means to assist in the detection of malicious bots on social media.

Notes

1 Here, we use the naming convention adopted by the IEEE. A/IS include, but are not limited to, machine learning, artificial intelligence, and neural networks approaches. Similarity to the initialism for adaptive instructional systems (AIS) is unfortunate, but unavoidable at this point.
2 Researchers initially believed that Game Theory provided a descriptive model. Many authors describe this approach in this manner.

References

Allhoff, F., Henschke, A., & Strawser, B. J. (2016). *Binary bullets: The ethics of cyberwarfare*. Oxford University Press.
Arquilla, J. (2013). Twenty years of cyberwar. *Journal of Military Ethics, 12*, 80–87.
Arquilla, J. & Ronfeldt, D. (1993). Cyberwar is coming! *Comparative Strategy, 12*, 141–165.
Axelrod, R. (2014). A repertory of cyber analogies. In E. Goldman & J. Arquilla (eds.), *Cyber analogies* (pp. 108–116). Naval Postgraduate School.
Balliet, D., Tybur, J. M., & Van Lange, P. A. (2017). Functional interdependence theory: An evolutionary account of social situations. *Personality and Social Psychology Review, 21*, 361–388.
Berkeley, J., Dietvorst, J. P. S., & Massey, C. (2015). Algorithm aversion: People erroneously avoid algorithms after seeing them err. *Journal of Experimental Psychology, 144*, 114–126.
Beskow, D. M., & Carley, K. M. (2019). Social cybersecurity: An Emerging national security requirement. *Military Review, 99*, 117.
Betz, D. J. (2017). *Cyberspace and the state: Towards a strategy for cyber-power*. Routledge.
Betz, D. J., & Stevens, T. (2013). Analogical reasoning and cyber security. *Security Dialogue, 44*, 147–164.
Blanchette, I., & Dunbar, K. (1997). Constraints underlying analogy use in a real-world context: Politics. In M. G. Shafto & P. Langley (eds.), *Proceedings of the Nineteenth Annual Conference of the Cognitive Science Society* (p. 867). Erlbaum.
Briscoe, E. J., Appling, D. S., & Hayes, H. (2014). Cues to deception in social media communications. In *Proceedings of the 47th Hawaii International Conference on System Sciences* (pp. 1435–1443). IEEE.
Brown, D. E., & Clement, J. (1989). Overcoming misconceptions via analogical reasoning: Abstract transfer versus explanatory model construction. *Instructional Science, 18*, 237–261.

Burleigh, T. J., & Schoenherr, J. R. (2014). A reappraisal of the uncanny valley: Categorical perception or frequency-based sensitization? *Frontiers in Psychology, 5*, 1488.

Burleigh, T. J., Schoenherr, J. R., & Lacroix, G. L. (2013). Does the uncanny valley exist? An empirical test of the relationship between eeriness and the human likeness of digitally created faces. *Computers in Human Behaviour, 29*, 759–771. https://doi.org/10.1016/j.chb.2012.11.021.

Burnes, B., & Cooke, B. (2013). Kurt Lewin's field theory: A review and re-evaluation. *International Journal of Management Reviews, 15*, 408–425.

Burstein, M. H. (1983). Concept formation by incremental analogical reasoning and debugging. In *Proceedings of the International Machine Learning Workshop* (pp. 19–25). Allerton House University of Illinois at Urbana-Champaign.

Cheetham, M., Suter, P., & Jäncke, L. (2011). The human likeness dimension of the "uncanny valley hypothesis": Behavioral and functional MRI findings. *Frontiers in Human Neuroscience, 5*, 126. https://doi.org/10.3389/fnhum.2011.00126.

Chen, Z., Mo, L., & Honomichl, R. (2004). Having the memory of an elephant: Long-term retrieval and the use of analogues in problem solving. *Journal of Experimental Psychology: General, 133*, 415.

Christensen, J. (1999). Bracing for guerrilla warfare in cyberspace. *CNN Interactive XXXX*. Retrieved November 19, 2019. https://cyber.harvard.edu/eon/ei/elabs/security/cyberterror.htm.

Chu, Z. Gianvecchio, S., Wang, H., & Jajodia, S. (2010). Who is tweeting on Twitter: Human, bot, or cyborg? In *Proceedings of the 26th Annual Computer Security Applications Conference* (pp. 21–30). Austin, Texas, USA.

Clement, C. A., Kurland, D. M., Mawby, R., & Pea, R. D. (1986). Analogical reasoning and computer programming. *Journal of Educational Computing Research, 2*, 473–486.

de Matos Alves, A. (2015). Between the "battlefield" metaphor and promises of generativity: Contrasting discourses on cyber conflict. *Canadian Journal of Communication, 40*, 389–405.

Dunbar, K. (1997). How scientists think: On-line creativity and conceptual change in science. In T. B. Ward, S. M. Smith, & S. Vaid (eds.), *Conceptual structures and processes: Emergence, discovery, and change* (pp. 461–493). American Psychological Association.

Dunbar, K. (2001). The analogical paradox: Why analogy is so easy in naturalistic settings yet so difficult in the psychological laboratory. In D. Gentner, K. Holyoak, & B. Kokinov (eds.), *The Analogical Mind: Perspectives from Cognitive Science* (pp. 313–334). MIT Press.

Efthimion, P. G., Payne, S., & Proferes, N. (2018). Supervised machine learning bot detection techniques to identify social twitter bots. *SMU Data Science Review, 1*, 5.

Feily, M., Shahrestani, A., & Ramadass, S. (2009, June). A survey of botnet and botnet detection. In *2009 Third International Conference on Emerging Security Information, Systems and Technologies* (pp. 268–273). IEEE.

Ferrara, E., Wang, W. Q., Varol, O., Flammini, A., & Galstyan, A. (2016). Predicting online extremism, content adopters, and interaction reciprocity. In *International Conference on Social Informatics* (pp. 22–39). Springer.

Fiske, A. P. (1992). The four elementary forms of sociality: Framework for a unified theory of social relations. *Psychological Review, 99*, 689–723.

Fiske, A. P., & Rai, T. S. (2014). *Virtuous violence: Hurting and killing to create, sustain, end, and honor social relationships.* Cambridge University Press.

Fletcher, G. J. (1983). The Analysis of verbal explanations for marital separation: Implications for attribution theory. *Journal of Applied Social Psychology, 13*, 245–258.

Friedkin, N. E. (1980). A test of structural features of Granovetter's strength of weak ties theory. *Social Networks, 2*, 411–422.

Gibson, W. (1984). *Neuromancer.* Ace.

Gick, M. L., & Holyoak, K. J. (1980). Analogical problem solving. *Cognitive Psychology, 12*(3), 306–355.

Gilbert, E., Karahalios, K., & Sandvig, C. (2008, April). The Network in the garden: An empirical analysis of social media in rural life. In *Proceedings of the SIGCHI Conference on Human Factors in Computing Systems* (pp. 1603–1612).

Gilbert, E., & Karahalios, K. (2009, April). Predicting tie strength with social media. In *Proceedings of the SIGCHI Conference on Human Factors in Computing Systems* (pp. 211–220). Boston, Massachusetts, USA.

Goldman, E. O., & Arquilla, J. (2014). *Cyber analogies.* Naval Postgraduate School.

Granovetter, M. S. (1973). The Strength of weak ties. *The American Journal of Sociology, 78*, 1360–1380.

Grinberg, M., Hristova, E., & Borisova, M. (2012). Cooperation in prisoner's dilemma game: Influence of social relations. In *Proceedings of the Annual Meeting of the Cognitive Science Society* (Vol. 34, No. 34, pp. 408–413). Sapporo, Japan.

Halevy, N., Chou, E. Y., & Murnighan, K. (2012). Mind games: The mental representation of conflict. *Journal of Personality and Social Psychology, 102*, 132–148.

Harrison, F., Sciberras, J., & James, R. (2011) Strength of social tie predicts cooperative investment in a human social network. *PLoS One, 6*, e18338.

Hoffman, R. R. (1980). Metaphor in science. In Richard P. Honeck & Robert R. Hoffman (eds.), *Cognition and figurative language* (pp. 393–423). Routledge.

Ishowo-Oloko, F., Bonnefon, J., Soroye, Z., Crandall, J., Rahwan, I., & Rahwan, T. (2019). Behavioural evidence for a transparency–efficiency tradeoff in human–machine cooperation. *Nature Machine Intelligence, 1*, 517–521.

Jiang, M., Cui, P., & Faloutsos, C. (2016). Suspicious behavior detection: Current trends and future directions. *IEEE Intelligent Systems, 31*, 31–39.

Jordan, T. (1999). *Cyberpower: The culture and politics of cyberspace and the Internet.* Psychology Press.

Kar, D., Nguyen, T. H., Fang, F., Brown, M., Sinha, A., Tambe, M., & Jiang, A. X. (2018). In T. Başar & G. Zaccour (eds), Trends and applications in Stackelberg security games. In *Handbook of dynamic game theory* (pp. 1–47). Springer.

Karim, A., Salleh, R. B., Shiraz, M., Shah, S. A. A., Awan, I., & Anuar, N. B. (2014). Botnet detection techniques: Review, future trends, and issues. *Journal of Zhejiang University Science, 15*, 943–983.

Kelley, H. H. (1967). Attribution theory in social psychology. *Nebraska Symposium on Motivation, 15*, 192–238.

Kelley, H. H. (1973). The Process of causal attribution. *American Psychologist*, 28, 107–128.
Kelley, H. H. (1980). Attribution theory and research. *Annual Review of Psychology*, 31, 457–501.
Kelley, H. H., Holmes, J. G., Kerr, N., Reis, H., Rusbult, C., & Van Lange, P. A. (2003). *An atlas of interpersonal situations*. Cambridge University Press.
Kelley, H. H., & Thibaut, J. W. (1978). *Interpersonal relations: A theory of interdependence*. John Wiley.
Kiesler, S., Sproull, L., & Miller, J. (1996). A prisoner's dilemma experiment on cooperation with people and human-like computers. *Journal Personality Social Psychology*, 70, 47–65.
Kroll, A. (2019). We're not ready for a massive digital terror attack. *Rolling Stone*. https://www.rollingstone.com/politics/politics-features/cambridge-analytica-christchurch-trump-snowden-brad-smith-881314/.
Lauland, A. (2016). Rather than fearing 'Cyber 9/11', prepare for 'Cyber Katrina'. *RAND*. https://www.rand.org/blog/2016/03/rather-than-fearing-cyber-911-prepare-for-cyber-katrina.html.
Lee, W. (1971). *Decision theory and human behavior*. Wiley.
Lewin, K. (1939). Field theory and experiment in social psychology. *American Journal of Sociology*, 44, 868–896.
Lewis, P. T. (1995). A Naturalistic test of two fundamental propositions: Correspondence bias and the actor-observer hypothesis. *Journal of Personality*, 63, 87–111.
Liberman, V., Samuels, S. M., & Ross, L. (2004). The name of the game: Predictive reputations versus situational labels in determining prisoner's dilemma game moves. *Personality and Social Psychology Bulletin*, 30, 1175–1185.
Liles, S. (2010). Cyber warfare: As a form of low-intensity conflict and insurgency. In C. Czosseck & K. Podins (eds), *Conference on Cyber Conflict Proceedings* (pp. 47–58). Tallinn, Estonia.
Liles, S., & Rogers, M. (2008, March). Cyber warfare as a form of low intensity conflict. In *Proceedings of the 9th Annual Information Security Symposium* (p. 1). West Lafayette, IN.
Lin, N., Dayton, P. W, & Greenwald, P. (1978). Analyzing the instrumental use of relations in the context of social structure. *Sociological Methods Research*, 7, 149–166.
Luce, R. D., & Suppes, P. (1965). Preference, utility, and subjective probability. In R. D. Luce, R. R. Bush, & E. Galanter (Eds.) *Handbook of mathematical psychology*, Vol. III. New York: Wiley. pp. 252–410.
Martin, J., Gonzalez, C., Juvina, I., & Lebiere, C. (2014). A description – experience gap in social interactions: information about interdependence and its effects on cooperation. *Journal of Behavioral Decision Making*, 27, 349–362.
Merritt, T., & McGee, K. (2012). Protecting artificial team-mates: More seems like less. In *Proceedings of the SIGCHI Conference on Human Factors in Computing Systems* (pp. 2793–2802). Austin, Texas, USA.
Messick, D. M. (1999). Alternative logics for decision making in social settings. *Journal of Economic Behavior & Organization*, 39, 11–28.
Myers, L. (2011). Do we need a cyber-Chernobyl? *IT News*. https://www.itnews.com.au/feature/do-we-need-a-cyber-chernobyl-261987.
Myerson, R. B. (1997). *Game theory*. Harvard University Press.

Nash, J. F. (1950). Equilibrium points in n-person games. In *Proceedings of the National Academy of Sciences of the United States of America* (Vol. 36, No. 1., pp. 48–49).

Neustadt, R., & May, E. (1986). *Thinking in time: The uses of history for policy makers*. Simon and Schuster.

Nguyen, R. (2013). Navigating jus ad bellum in the age of cyber warfare. *California Law Review, 101*, 1079–1129.

Nye, J. S. (2010). *Cyber power*. Harvard University, Belfer Center for Science and International Affairs.

Nye, J. S. (2011). Nuclear lessons for cyber security? *Strategic Studies Quarterly, 5*, 18–38.

Oudah, M., Babushkin, V., Chenlinangjia, T. & Crandall, J. W. (2015). Learning to interact with a human partner. In *Proceedings of the Tenth Annual ACM/IEEE International Conference on Human-Robot Interaction* (pp. 311–318). ACM.

Passer, M. W., Kelley, H. H., & Michela, J. L. (1978). Multidimensional scaling of the causes for negative interpersonal behavior. *Journal of Personality and Social Psychology, 36*, 951–962.

Pea, R. D., & Kurland, D. M. (1984). On the cognitive effects of learning computer programming. *New Ideas in Psychology, 2*, 137–168.

Perkovich, G., & Levite, A. E. (2017). *Understanding cyber conflict: Fourteen analogies*. Georgetown University Press.

Pillutla, M. M., & Chen, X.-P. (1999). Social norms and cooperation in social dilemmas: The effects of context and feedback. *Organizational Behavior and Human Decision Processes, 78*, 81–103.

Postmes, T., & Spears, R. (1998). Deindividuation and antinormative behaviour: A meta-analysis. *Psychological Bulletin, 123*, 238.

Rai, T., & Fiske, A. (2011). Moral psychology is relationship regulation: Moral motives for unity, hierarchy, equality, and proportionality. *Psychological Review, 118*, 57.

Rapoport, A., & Chammah, A. M. (1965). *Prisoner's dilemma: A study in conflict and cooperation*. University of Michigan Press.

Rapoport, A., & Guyer, M. (1966). A taxonomy of 2 × 2 games. *General Systems, 11*, 203–214.

Rattray, G. J. (2009). An Environmental approach to understanding cyberpower. *Cyberpower and National Security, 253*, 274.

Rumelhart, D. E., & Abrahamson, A. A. (1973). A Model for analogical reasoning. *Cognitive Psychology, 5*, 1–28.

Schoenherr, J. R. (2022). *Trust in the age of entanglement: Designing ethical artificial intelligence from popular science to cognitive science*. Routledge.

Schoenherr, J. R. (forthcoming). Learning engineering is ethical. In J. D. Goodell (ed.), *Learning engineering toolkit*. Routledge.

Schoenherr, J. R., & Burleigh, T. J. (2020). Dissociating affective and cognitive dimensions of uncertainty by altering regulatory focus. *Acta Psychologica, 205*, 103017.

Schoenherr, J. R., & Thomson, R. (2020). Dissociating cognitive and affective uncertainty using a general linear classifier. In J. Schoenherr (ed.), *Fechner day* (pp. 20–26). Akdeniz University.

Sedjelmaci, H., Brahmi, I. H., Boudguiga, A., & Klaudel, W. (2018, January). A generic cyber defense scheme based on Stackelberg game for vehicular network. In *2018 15th IEEE Annual Consumer Communications & Networking Conference (CCNC)* (pp. 1–6). IEEE.

Shao, C., Ciampaglia, G. L., Varol, O., Yang, K. C., Flammini, A., & Menczer, F. (2016). The Spread of low-credibility content by social bots. *Nature Communications, 9*, 1–9.

Shi, X., Adamic, L. A., & Strauss, M. J. (2007). Networks of strong ties. *Physica A: Statistical Mechanics and its Applications, 378*(1), 33–47.

Sjouwerman, S. (2020). Ransomware Incidents Increase 131 Percent with the SMB Being the Primary Target. Retrieved from https://blog.knowbe4.com/ransomware-incidents-increase-131-percent-with-the-smb-being-the-primary-target

Spinello, R. A., & Tavani, H. T. (2005). *Readings in cyberethics*. Jones & Bartlett Learning.

Sproull, L., & Kiesler, S. (1991). Computers, networks and work. *Scientific American, 265*(3), 116–127.

Sternberg, R. J. (1977). Component processes in analogical reasoning. *Psychological Review, 84*, 353.

Subrahmanian, V., Azaria, A., Durst, S., Kagan, V., Galstyan, A., Lerman, K., Zhu, L., Ferrera, E., Flammini, A., & Menczer, F. (2016). The DARPA Twitter bot challenge. *IEEE Computer, 6*, 38–46.

Sulek, D., Moran, N., & Principal, B. A. H. (2009). What analogies can tell us about the future of cybersecurity. *The Virtual Battlefield: Perspectives on Cyber Warfare, 3*, 118.

Tambe, M. (2012). *Security and game theory: Algorithms, deployed systems, and lessons learned*. Cambridge University Press.

Tenbrunsel, A. E., & Messick, D. M. (1999). Sanctioning systems, decision frames, and cooperation. *Administrative Science Quarterly, 44*, 684–707.

Thomas, T. L. (2006). *Cyber mobilization: A growing counterinsurgency campaign*. Foreign Military Studies Office (Army) Fort Leavenworth Ks.

Van Haaster, J., Gevers, R., & Sprengers, M. (2016). *Cyber guerilla*. Syngress.

Varol, O., Ferrara, E., Davis, C., Menczer, F., & Flammini, A. (2017, May). Online human-bot interactions: Detection, estimation, and characterization. In *Proceedings of the international AAAI conference on web and social media* (Vol. 11, No. 1). Montreal, QC, Canada.

von Neumann, J., & Morgenstein, O. (1947). *Theory of games and economic behavior*. Princeton University Press.

Von Stackelberg, H. (1934). *Marktform und Gleichgewicht*. Springer.

Vosniadou, S., & Ortony, A. (1989). *Similarity and analogical reasoning*. Cambridge University Press.

Wagner, C., Mitter, S., Körner, S., & Strohmaier, M. (2012). When social bots attack: Modeling susceptibility of users in online social networks. In *Proceedings of the 21st International Conference on the World Wide Web* (pp. 41–48). Lyon, France.

Wald, R., Khoshgoftaar, T. M., Napolitano, A., & Sumner, C. (2013). Predicting susceptibility to social bots on Twitter. In *Proceedings of the 14th IEEE*

International Conference on Information Reuse and Integration (pp. 6–13). IEEE.

Wulff, D., Mergenthaler-Canseco, M., & Hertwig, R. (2018). A meta-analytic review of two modes of learning and the description-experience gap, *Psychological Bulletin*, Vol. 144, pp. 140–176.

Xu, B., Liu, L., & You, W. J. (2011) Importance of tie strengths in the prisoner's dilemma game on social networks. *Physics Letters A, 375*, 2269–2273.

Yin, A., Korzhyk, D., Kiekintveld, C., Contizer, C., & Tambe, M. (2010). Stackelberg vs. Nash in Security Games: interchangability, Equivalence, and Uniqueness in *Proceedings of the International Conference on Autonomous Agents and Multiagent Systems* (AAMAS).

Zeidanloo, H. R. & Manaf, A. A. (2009). Botnet command and control mechanisms. In *Proceedings of the Second International Conference on Computer and Electrical Engineering* (pp. 564–568). IEEE.

Zeidanloo, H. R., Shooshtari, M. J. Z., Amoli, P. V., Safari, M., & Zamani, M. (2010, July). A Taxonomy of botnet detection techniques. In *2010 3rd International Conference on Computer Science and Information Technology* (Vol. 2, pp. 158–162). IEEE.

Zhu, M., & Martínez, S. (2011). Stackelberg-game analysis of correlated attacks in cyber-physical systems. In *Proceedings of the 2011 American Control Conference* (pp. 4063–4068). doi:10.1109/ACC.2011.5991463.

Chapter 9

Deep Blue Wants You

Identifying and Addressing Sources of Bias in AI Systems to Support Human Resources Decisions

Aryn Pyke, F. Jordan Richard Schoenherr, and Robert Thomson

1 Introduction: The Human Resources Use Case for AI

According to Tambe et al. (2019), "AI [Artificial Intelligence] conventionally refers to a broad class of technologies that allow a computer to perform tasks that normally require human cognition, including adaptive decision making". For some, advancements in AI have introduced concern about how labor will be reallocated from humans to AI algorithms in the modern workforce while also raising fears about the potential loss of human jobs (Su, 2018). For those with such concerns, it might throw gas on the fire to learn that AI and machine learning algorithms might play an increasing role in the human resources (HR) processes that determine which humans are selected for the remaining jobs. We suggest, however, that allowing AI to play a role in HR actually has the potential to introduce a number of important benefits, including more unbiased, and therefore potentially more ethical, selection processes (Kleinberg et al., 2018b). In a survey of thousands of recruiters and hiring managers worldwide, 43% reported the view that a benefit of AI in HR is to remove human bias (LinkedIn, 2018a). The question becomes, how can we increase the probability that AI HR systems will live up to this potential for less biased selections? Furthermore, how will we readily and reliably detect when an AI HR system has failed to do so (Silberg & Manyika, 2019)? For the purposes of brevity in this chapter, we will use the term "AI" (Artificial Intelligence) as an umbrella term to include any computational algorithms or machine learning systems to support HR decision making.

Given the diversity and increasing use of AI systems, the application of AI to HR serves as a useful case study to provide insight into social and ethical issues that are generally relevant to all AI systems involved in decisions about whether or not an individual will be granted an important

DOI: 10.4324/9781003030928-13

opportunity. Kleinberg et al. (2018b) refer to such decisions as *screening decisions*. This category of possible AI applications would include the selection of which individuals will be: (i) recruited, hired, and promoted; (ii) granted loans and mortgages; (iii) admitted to college; (iv) granted particular government benefits; (v) given access to good health insurance rates and medical treatments; (vi) granted bail or parole in the justice system; etc. Due diligence must be devoted to examining the ethical ramifications of these processes and their outcomes because decisions of whether or not individuals (or groups) are selected for such opportunities can have a huge impact on their quality of life.

In this chapter, we first give a brief overview of some types of AI systems already in use in HR, and then discuss the benefits that the AI could contribute to HR. Finally, we discuss potential sources of bias in AI selection systems, which we have grouped into three broad categories: (i) Bias in AI developers and system design; (ii) Bias in the data used to train AI systems; and (iii) Bias in use of the AI systems. Within each category we suggest and/or review possible solutions to reduce the bias.

1.1 Examples of AI Systems Already in Use in HR

According to a report from LinkedIn (2018b), about 22% of organizations are using AI or other analytics in HR. Within the HR domain, organizations develop and execute procedures to recruit, select, retain, and promote employees that will fit and perform well within their organization. Captain (2016) touches on several aspects of HR in which AI systems are starting to play an increasingly important role. In terms of recruitment, *Talent Bin* is a system associated with Monster.com that uses social media data to proactively find a match between jobs and possible recruits. To assess candidates' job skills (e.g., knowledge of Microsoft Excel), the *Interviewed* system can automatically administer and score various tests. Another example is the *RoundPegg* system, which strives to assess the "cultural fit" of a candidate for an organization based on how they categorize words or phrases (e.g., "fairness" and "being team oriented") into their nine most important and nine least important values. RoundPegg compares the candidates' categorization patterns with those collected (and presumably aggregated) from existing employees in the organization. *HireVue* is a system that specializes in the analysis of video interviews, and strives to assess personality attributes like empathy, motivation, and engagement from data like the candidate's word choice, rate of speech, and facial expressions.

In terms of professional development for existing employees, there are also AI systems (including one allegedly used by IBM, Tambe et al., 2019) that can recommend and customize training opportunities that are matched to an employee's needs (Barnard, 2019). There are also AI systems, like IBM's *Blue Match* software and a product by *Quine*, that can suggest career advancement moves (e.g., job openings the employee might want to apply for within an organization; Tambe et al., 2019). In 2018, 27% of IBM employees who received a promotion or new job were assisted by the Blue Match system (Rosenbaum, 2019). Although the examples of systems mentioned above might apply to different aspects of the HR process, a common function of these types of systems is that they are all involved in the selection of which individuals will have access to (or at least be made aware of) certain opportunities.

2 The Benefits AI Might Bring to HR Processes

Human resource selection has shifted away from *ad hoc* selection based on interpersonal accord, to unstructured interviews, to structured interviews and psychometrically validated assessment instruments. However, despite innovations, inequality remains a persistent concern. For instance, referencing several metrics (e.g., the executive parity index, EPI), Gee (2018) notes that there remain large disparities between racial and gender categories in terms of who occupies white-collar positions. Given this room for improvement, the introduction of AI selection systems into the HR process might be capable of reducing biases in a way that humans have not.

At a meta-level of analysis, AI systems can be applied to the problem of detecting bias in selection patterns (Silberg & Manyika, 2019). Further, the attempt to create and introduce AI systems to support screening decisions might already be promoting fairer selection systems by calling attention to, and promoting, increased scrutiny of potential bias. For example, Silberg and Manyika (2019) suggest that a bias in the human selection processes for admitting candidates to a British medical school (against applicants with female and non-European names) likely only came to light because they introduced an AI system that mimicked these established biases. The use of an AI prompted closer scrutiny of the selection patterns. In cases where an AI selection system might be found to have bias, the kicker is that such bias is typically a reflection of human selection precedents used to train the system. Notably, computers do not inherently harbor emotional biases or prejudices. Thus, as we will discuss later in the chapter, if developers can avoid introducing such bias, AI selection systems might help avoid the application of personal and societal biases.

At a lower level of analysis, features such as speed, scalability, and reductions in costs are typically cited as key beneficial changes to organizations if they adopt AI processes into their workflow (Aspan, 2020; LinkedIn Global Recruiting Trends, 2018). The types of benefits (or perils) that of interest in this chapter, however, are those that could impact bias and fairness in the HR process of selecting individuals for opportunities.

Interestingly, one of the advantages of AI systems is that they make decisions potentially thousands of times faster than humans. For instance, Google Search returns results for your keyword search in a fraction of a second while using input from billions of websites. Such speed can not only improve efficiency but could also contribute to reducing bias. Organizations face limitations of time and evidence (Simon, 1997), so when a large number of candidates apply for an opportunity, vetting each candidate "by hand" becomes increasingly untenable, shallow, and can threaten the validity of the process. For example, Google receives two million applications a year (Schneider, 2017). To vet such a large number of applicants by hand, Schneider (2017) suggests that recruiters might make a decision to reject a candidate after investing only about six seconds looking at his or her resume. He also reports that one of the six pieces of information that recruiters tend to focus on in these six seconds is the candidate's name. Notably, names provide cues about sex, race, ethnicity, and religion, and research indicates that employers are vulnerable to name-related biases. For example, when fictitious resumes with White-sounding versus African-American-sounding candidate names were submitted to job ads, the former received 50% more callbacks for interviews (Bertrand & Mullainathan, 2004). Thus, if thousands or millions of résumés must be vetted by hand, the minimal human time available to vet each résumé, and the amount of that limited time spent on a cue that might be processed according to biased stereotypes is quite worrisome. Cursory human involvement of this kind does not seem to ensure a personal touch. Candidates would seem better served by an AI selection (or at least filtering) algorithm that is very fast but can nonetheless incorporate and integrate more information into the decision process.

Notably, AI cannot only increase the number of candidates that can be processed in a given time, but can also make the automated processing of far more information about each candidate feasible, from a variety of sources and in a variety of formats. For example, instead of limiting the analysis to résumés, an AI system could also process data from an individual's social media (e.g., TalentBin), work samples, answers on assessment tests to evaluate skills or personality characteristics (e.g., Interviewed and Koru), and interview videos (e.g., rate of speech, word choice, and facial expressions in the HireVue system). To the extent that more complete data might reduce the uncertainty and guesswork involved in a decision, the ability of an AI system to process more data has the potential to increase the fairness of the selection process.

Furthermore, if an AI can do a deeper-dive evaluation of a candidate in one pass, this might also improve the accuracy of the selection process. Evidence shows that in a human decision-making system, as the number of selection stages increases, the probability of selecting the "best" candidate decreases. Consider a three-stage candidate selection process that includes: a headhunter identifying candidates, a working group selecting from among the submitted résumés, and a management team making the final hiring decision. If the reliability of each assessment stage is relatively high (e.g., 0.8), the probability that the best candidate is selected is only about 50% ($p = 0.5^3$). The more biases that are introduced at each given level, the less likely the best candidate will be selected (i.e., the principle of invidious selection; Thorngate, 1988; Thorngate & Carroll, 1990).

3 The Potential Sources of Bias in AI Selection Systems

Though we are optimistic about the potential benefits of AI to support ethical HR processes, we must also be cognizant of the potential dangers. AI systems do not inherently harbor emotional biases, prejudices, or social agendas, and, in theory, and can be prevented from explicitly knowing the gender, religion, skin color, age, etc., of job candidates. Bias, however, can seep in due to human influence in the design, training data, and use of AI systems. In subsections below, we will discuss each of these potential sources of bias and suggest or review ways in which they might be remediated.

First, however, we want to acknowledge that bias in an AI selection system is particularly worrisome if the AI system might ultimately replace a group of HR decision makers. Within a human hiring committee, the multiple perspectives can provide some checks and balances against the biases of individual members. A unitary AI system is not subject to such internal checks and balances. The scope of this issue is magnified by the fact that a particular AI selection system might not just influence decisions within a single organization. Rather, companies developing HR AI systems are offering them as HR services for many organizations. For example, *Koru*, an AI system that can evaluate candidates based on responses to an on-line assessment, has large clients including REI, Zillow, Yelp, Airbnb, Facebook, LinkedIN, Reebok, and McKinsey & Company. Similarly, RoundPegg, the AI system to assess cultural fit in a company, has clients including Experian, ExxonMobil, Razorfish, and Xerox (Captain, 2016). Note that we are not evaluating or casting aspersions on these particular systems—our point is simply that *if* there is bias in a single AI-based selection system, a vast number of people could be impacted.

As we mentioned, bias in AI systems can stem from human influence in development and design, training data, and use/interpretation of AI systems. To clarify these possible loci of vulnerability to bias, we will briefly describe some properties of the underlying technology. Many computer decision-making systems are based on machine learning algorithms. As the name suggests, the algorithm itself learns from data to which it has access, just as a child learns from the available information in her/his environment. Clearly if the information available to the AI (or child) is incorrect, incomplete, or biased, this will impact what is learned. For example, in the historical data set provided to the system, perhaps all successful candidates or "good" workers were male. Thus, as will be discussed in more detail below, the training data can introduce bias (Munoz et al., 2016; Raub, 2018), potentially perpetuating existing systems of bias that produced the data.

Given that the machine has no a priori prejudices and is self-taught, one might be inclined to think that the data sets selected to train the system might be the only potential source of bias. However, beyond selecting which data sets the systems learn from, AI developers also design and constrain the system in a way that determines which fields (types of information) the algorithm can "see" from these data sets. Most AI systems (unlike children) do not have eyes and ears of their own, so designers must decide a priori what types of input information are relevant/available to the AI decision making process. As an overly-simplistic example, an AI system might be designed to expect (or at least accept) the sex, race, and religion of candidates as inputs.

Another potential vulnerability to bias can occur when the recommendations provided by an AI system are processed and acted upon by human decision makers. Some individuals might be biased to blindly defer to the recommendations of an automated system (i.e., Automation bias; Robinette et al., 2016), and fail to notice indications of bias. Others might take a potentially unbiased ranked list of recommendations from the system and accept or reject some recommendations based on personal prejudices related to factors such as race or gender if the human decision maker might have access to that information even if the AI system did not.

This rudimentary introduction provides some context for the claim that bias might be introduced in the design, training data, and interpretation of AI systems. We now discuss these issues in more detail.

3.1 Bias in AI Developers and Design

Human AI developers are obviously vulnerable to prejudices and biases, which might even be non-conscious (Staats et al., 2016). As touched on above, developer biases can influence not only which data sets are chosen to train the system, but also aspects of system design which determine which fields (types of information) the algorithm can "see" from these data sets (e.g., name and/or sex of candidate). Developers also might be involved in determining the "resolution" of the information that reaches the system. For example, one might provide a system with a candidate's age to the nearest year, or one might provide a 2-level distinction ("young" if under 30 and "old" if over 31). Crucially, those who are not adept at noticing possible manifestations of bias might be unaware of how aspects of the design might lead to discrimination. Below we review several possible approaches to reduce bias in AI systems that might be introduced by possibly unconscious bias in developers and system design decisions.

Increase Developer Diversity. The ability to detect manifestations of bias can be developed via first-hand experience at being subjected to discrimination, e.g., members of minority groups and women. Currently, however, women and many minority groups are dramatically underrepresented in computing fields (Mundy, 2017). For example, Mundy (2017) reported that only 25% of computing and mathematical jobs are held by women. Similarly, Tiku (2018) reports that Google's workforce is 53% white, 36% Asian, 3% black, 4% Hispanic/Latinx, and 4% were classified as multiracial.

As such, some have suggested that bias in AI systems could be reduced by having a larger proportion of women and minorities in the AI developer community (Barnard, 2019; Gebru, 2019; Silberg & Manyika, 2019). Thus, the checks and balances inherent in a diverse human hiring committee could be moved upstream by having a diverse development team for the AI system. This approach would certainly help to resolve an imbalance in this sector of the workforce, and could serve to raise awareness of possible manifestations of bias in AI design within the AI community as a whole. This approach reflects the idea of *compositional fairness*: if the group making a decision (here, AI design decisions) contains a diversity of viewpoints, then the outcome will be deemed fair (Silberg & Manyika, 2019). Each individual AI development project, however, would not be guaranteed of having a diverse team of developers. Furthermore, AI developers who happen to be minorities and women will not all necessarily be focused on or adept at detecting sources of bias, nor might they all welcome that "gatekeeping" function as part of their implicit (or explicit) role on the development team. Nonetheless, we argue that increasing diversity among developers would certainly be a step in the right direction.

Train AI Developers to Detect Bias. It has also been suggested that AI developers should all receive training to make them more adept at recognizing potential sources of bias in their designs and training data, and that the industry should perhaps require AI developers to have certification of such training (Miller et al., 2018). Human organizational developers already receive training to identify manifestations of bias, oppression, and discrimination (Miller et al., 2018), but developers of HR AI systems, which might increasingly make decisions that impact human lives, do not (Mundy, 2017). Even awareness of bias, however, might not induce developers to correct it if the consequences of such bias are unclear. Katz, Miller, and Gans (2018) suggest that developers also be informed about the value-added and return-on-investment of inclusive hiring practices, and about the costs of biased practices to organizations in terms of poor public relations and lawsuits.

Though we agree that increasing the ability of individuals (including AI developers) to recognize manifestations of bias is a laudable goal in general, we do not think that requiring all AI developers to have such training is necessarily an optimal solution. AI developers might be involved in designing a wide variety of applications rather than necessarily focused on developing AI for a specific domain (like HR). Factors of prejudice and discrimination might not be the key design concerns in many projects—for example, in the design of an AI to sort glass from plastic recyclables. For projects that do involve making decisions (like HR decisions) about which individuals will be selected for an opportunity and which will not, it would seem to make more sense to capitalize on existing experts trained in bias-detection and have an interdisciplinary development team. Thus, we recommend that an Organizational Development / Human Resources professional be added to the development team for such projects to check for possible bias in the types of input information the AI has access to (e.g., applicant hometown) and the training data set (e.g., a set where an employee's quality rating was made only by white males).

Increase the Transparency of AI Systems Themselves: "Explainable AI". Employers are legally accountable for making decisions in a fair manner (Tambe et al., 2019). However, to avoid using a selection process characterized by bias (and/or to correct bias in a system), it is first necessary to detect that bias is present. Unfortunately, bias is not always easy to detect or prove in either human or AI selection processes. Human decision-makers might not be fully aware of their own biases and cognitive processes (Staats et al., 2016) and/or they might misrepresent themselves and lie, especially if faced with the possibility of legal repercussions (Kleinberg et al., 2018b). One might expect that it would be easier to detect bias in a computational system, because, although it is not possible to directly observe human thought processes, one can observe the code that dictates how an algorithm works. There are two problems with this assumption, however. First, some AI algorithms are proprietary (Tambe et al., 2019), including those developed by Google and Amazon (Miller et al., 2018). Second, access to the code of an algorithm might not provide a clear understanding of how it works.

Recall that many of the algorithms used in this domain are machine learning algorithms. As such, the system learns and adjusts its internal properties in response to exposure to training data. These internal properties of the system are far from straightforward to interpret with respect to predicting the behavior of the system or characterizing the basis on which decisions are made. The high-level implications of these internal properties are often opaque even to the AI developers themselves (e.g., Adadi & Berrada, 2018; Knight, 2017; Thomson & Schoenherr, 2020). In the next two paragraphs, we attempt to briefly describe the operation of a system to provide some insight into how this can be the case, however, these paragraphs might be skipped if a technical description is not of interest.

Consider a case in which one type of information that the system gets as input is the college that a candidate attended. To be processed by the machine learning algorithm, this information must be coded as a numeric binary value (a series of 1s and 0s). It takes a code with a length of 13 binary digits (1s and 0s) to be able to distinctly represent each of the approximately 5,000 universities and colleges in the United States. One candidate's college could be represented as 1001110001000. This information is treated by the system as thirteen separate inputs (one per binary digit). Other types of inputs are similarly translated into binary representations. For example, the candidate's 7 years of work experience might be represented as 00111. When all the input information is represented this way and concatenated, the input to the system is a long series of binary digit inputs. At this point the system might not have clear knowledge of which of those binary digits "go together" in terms of representing different pieces of higher-level information (i.e., college attended, years of work experience, etc.).

The structure of the system can be considered analogous to a neural network wherein subsets of the binary digit inputs are combined and recombined in a large number of "arbitrary" ways—and each combination is represented by a node in the network. The inclusion of any particular input (binary digit) in a combination node is represented by a line (link) in the network from that input digit to that combination node. Each input node contributes to many of these combination nodes in the next layer of the network. The output of each combination node is also a binary digit whose value is determined by a weighted average of all the values of the node's inputs.

Each link (from input node to combination node) is associated with a value that determines the weight of that input digit in determining the weighted average (output) of the combination node. These numerical weights associated with the links to combination nodes are the internal properties of the system that get adjusted during training. Depending on the weights, which inputs effectively contribute to a given combination node might not even include all the digits used to represent a whole piece of high-level information. For example, a node might combine the third binary digit from the college input (0) and the fourth binary digit from the years of work experience input (1), and many other digits that, individually, are difficult to interpret. The weighted combination of these values which is the output of the combination node is even more difficult to interpret in high-level terms. The outputs from the different combinations might be further combined and recombined in a large number of ways by feeding into nodes in subsequent layers of the network. Ultimately, the final layer of combination nodes yields the output for the system, which might represent, say, the evaluation of the candidate out of 5. This decision might be represented by 3 binary digit output nodes (e.g., a rating of 4 = 100). Amid all these combinations and recombinations of binary digits that are not always kept together in their original groups (e.g., 00111 for 7 years of work experience), it is inherently difficult to determine exactly how an original high-level input contributed to the decision. The purpose of an AI selection system is to make a categorization, not to reveal how that decision was made in a way understandable or articulable by humans. In all, knowing the internal parameters of the system (i.e., the numerical weights of the links in the network) does not make the system's decision easy to predict or explain.

Because it can be inherently difficult to "unpack" the high-level (i.e., human interpretable) decision criteria being used by AI/machine learning systems, some efforts have been made in a research area called Explainable AI (XAI) that aims to increase the transparency of such systems (for overviews see Adadi & Berrada, 2018; Thomson & Schoenherr, 2020). One approach is to design AI systems to be more modular so that one could hypothetically provide certain types of information (e.g., work experience) to one decision module and other types of information (e.g., college attended) to another decision module so that you can tell which modules (types of information) are influencing the decision outputs (Core et al., 2006). Similarly, there are salience-based techniques that try to highlight the inputs that were most "active" in the AI system's output decision (Samek & Müller, 2019). Another approach is to develop other computational systems that take the properties of the AI system in question as inputs and that yield human-interpretable descriptions of the criteria as outputs (Ras et al., 2018; Ribeiro et al., 2016). The aforementioned techniques, however, are currently still under development and are years away from practical deployment. Furthermore, in many cases, the techniques would not scale to the large amounts of data seen in HR systems.

Although it might not always be feasible to peer into the black box of the system to directly ascertain its selection criteria, one can examine the input and output patterns of the system and determine if the overall behavior of the system meets or violates certain metrics. This is the topic of the next section.

Equip Developers with Appropriate Quality Assurance Metrics. Even if the inner workings of an AI system are somewhat opaque one can still test the output behavior of the system to see if it is operating in an acceptable way. A typical way to test a machine learning system is to divide the training data set into two parts and use one for training the system and one to test the system. When we input the "test" subset of the training data into the system, it is already known whether or not these individuals were hired and/or had high performance ratings. Thus, we can check to see if the system will rank these individuals "correctly"—i.e., according to how they were assessed by a human manager. If so, we expect that if we input new candidates into the system, it will rank them according to the precedents set by the human managers whose decisions are represented in the training set.

Ensuring a system behaves in a way that is consistent with the behavior of prior human managers, however, by no means assures that it is behaving "fairly", but it is not always obvious how "fairness" should be defined, measured, and tested. There might even be potential trade-offs between different kinds of fairness (e.g., individual vs. group) and between fairness and other objectives (Silberg & Manyaki, 2019). Several possible ways to quantitatively assess the fairness of machine learning output patterns are reviewed by Verma and Rubin (2018). An example of one possible fairness metric is *group fairness* (or *statistical parity*). A system meets this criterion if both protected and unprotected groups have equal probability of being assigned a good rating. Another possible measure is *false positive error rate balance* (*predictive equality*), which is related to the fact that we sometimes predict that someone will be a good performer and grant them an opportunity, but they fail to meet expectations. This criterion is satisfied if the system is equally likely to overestimate the performance of a non-protected group as a protected group.

A discussion of all possible fairness metrics and their trade-offs is beyond the scope of this chapter, but a key point is that the onus should not just be on the development team to determine the desired behavior of the system. Developers might spontaneously initiate some projects, but, while they are on the clock, they are presumably typically tasked by "higher-ups" with projects. In these circumstances, they should be provided with specifications as to intended functionality—to include metrics for testing whether the decision patterns of the AI conform to the organization's intentions, anti-discrimination laws, and emerging ethical standards for AI. A framework called *FATE* (Fairness, Accountability, Transparency, and Ethics in AI; e.g., Cathe, 2018; Greene et al., 2019; Jobin et al., 2019) has been uniting AI researchers to investigate and address such social and ethical impacts of AI.

If the system's raw output patterns do not conform to the desired functionality, there are post-processing techniques that can transform some of the system's ratings after they are made to better conform to an intended fairness metric (Hardt et al., 2016, cited in Silberg & Manyika, 2019).

3.2 Bias in Training Data

Based on some available information about a candidate, AI HR systems can function as categorizers. These systems can categorize applicants in terms of which jobs or work roles they seem suited for within a company, and they can categorize candidates' in terms of predicted level of fit and performance in a particular work role. For example, Amazon. com Inc.'s AI system rated candidates on a scale of 1–5 (Dastin, 2018). Such machine learning systems need a data set from which to learn how to perform these categorizations. These training data can include information about prior candidates and/or employees (e.g., from résumés and online social networks) together with some measure of their fit or performance (e.g., manager's performance ratings). The system then "learns" how to weigh and combine the types of information to which it has access to categorize candidates in accord with the precedents established by the training data set. Once trained, the system can intake information about new candidates and classify them into work roles and/or predicted levels of fit or performance using the same procedure.

Unfortunately, AIs trained on available HR datasets have been found to have bias in their predictions. For example, in 2015, Amazon's experimental AI system produced predicted ratings that were systematically lower for women than men in technical occupations (Dastin, 2018). The data used to train the algorithm over a prior ten-year period and success/performance ratings reflected the male dominance of the tech industry in that period. Thus, as Silberg and Manyika (2019) note: "the underlying data rather than the algorithm itself are most often the main source of the [bias]". Below we outline several potential issues of bias in HR training data and review or propose possible approaches to mitigate such issues.

Datasets Do Not Include All Types of Potentially Good Performers. HR Datasets are fundamentally incomplete in that they can only include performance ratings for people who applied, were hired, and were thus given an opportunity to perform and be rated. The AI system is supposed to learn about properties of individuals that predict good performance, but if training datasets do not include all individuals who would have been good performers, the AI might fail to learn about the properties that could identify these missing types of high-performing individuals—for example, those who were never given an opportunity to perform, potentially due to biased hiring practices. Recall that Gee (2018) reported large disparities between racial and gender categories in terms of who occupies white-collar positions. Thus, HR datasets for white-collar jobs might exclude many women and members of minority groups who were not given an opportunity to occupy these positions, but could well have excelled. Such datasets might include candidates who applied and were not hired, and might assign such candidates a default minimum performance rating—which might be worse than excluding them from the dataset altogether. Had some of these candidates been given the opportunity, they might have performed well. Nonetheless, if they are assigned a low "assumed" performance rating based on not being hired, the AI will inappropriately associate their attributes with poor fit and low performance.

The datasets also do not include individuals who did not apply—possibly in anticipation of a biased hiring process—but who might have nonetheless been hired and/or might have proved high performers. This can systematically skew the demographics of the datasets because there can be differences across groups in terms of job-application behavior. For example, men who meet only 60% of the qualifications listed for a job might nonetheless apply, whereas women typically only apply if they meet 100% of the listed qualifications according to an internal Hewlett Packard report. In practice, the qualifications listed on a job posting might not actually all be necessary to be able to do the job well and some skills could be rapidly acquired on the job. Accordingly, when people opt not to apply for jobs when they do not meet all the listed qualifications, the reason is rarely because they do not think they could perform well at the job (only 10% of women & 12% of men; Mohr, 2014). Rather, the decision not to apply was more commonly because of concerns about the hiring process (Mohr, 2014).

Indeed, the job description/qualifications that the employer chooses to list might introduce bias into the hiring process that is unrelated to the demands of the job. For example, some ads are explicitly or implicitly biased in terms of gender (Kuhn & Shen, 2013). Some qualifications in job ads might even simply serve the function of—somewhat arbitrarily—reducing the number of applications to be processed by the HR team. This reason for listing possibly extraneous qualifications in a job description (to make reviewing applications possible in the available time by a human hiring team) would not be as necessary if an AI was processing the applications. Thus, datasets of prior employees and applicants do not include examples of all types of individuals who could perform well at the job.

The problem of incomplete data can be understood by using a signal detection theory framework, which measures the ability of a system to identify a "signal" (here, a good candidate) using four metrics: hits, correct rejections, false alarms, and misses (Abdi, 2007). In the HR context, if a candidate is good and the trained system classifies her as good, that is a hit. If a candidate is poor and the trained system classifies her as poor that is a correct rejection. If a candidate is poor and the trained system classifies her as good, that is false alarm. And finally, the case germane to the current discussion is a miss, which occurs when a candidate is good, but the trained system classifies her as poor. As the AI was trained on a dataset that excluded (or assigned default low ratings for) potentially good candidates, it is ill-prepared to identify some types of good candidates when put into practice—i.e., it is susceptible to "misses".

The incompleteness of a dataset might be reflected in the database having lower numbers of women and members of minority groups represented in the set relative to the number of other candidates. For example, if the majority of the dataset consists of white men, it is also typically the case that white men will be the majority among the high performers within the set. Thus, when the AI learns which properties are associated with high performance, it might be biased to learn possibly incidental properties common to white men, because of the high exposure to data examples of that type. For example, if a child is mostly exposed to dogs of a particular type (e.g., chihuahuas) she might be inclined to think, incorrectly, that specific properties of that breed are representative of all dogs.

How might we mitigate this problem? One possible approach is to try to use more modern datasets which hopefully reflect more equitable hiring practices and representative demographic distributions. For example, in 1950 the share of women in the labor force was 30%, but it reached almost 47% by 2000 (Toossi, 2002). However, training an AI requires a substantial amount of data so limiting the data to relatively recent hiring history might not afford a dataset of sufficient size. Another possibility might be to try to use datasets from companies or regions that have had more long-standing equitable hiring practices. A potential wrinkle with that approach, however, is that the properties of people that predict success in other companies, industries or regions might not generalize to predict success and fit in the current company of interest.

Another approach is to try to modify or supplement the available dataset in a way that helps resolve imbalances. If the dataset is impoverished in terms of the number of individuals from a particular demographic (e.g., women) one could repeat those "rows" in the dataset when inputting the data to the AI, to emulate having a more equal proportion of women in the set. This technique is called *oversampling* and has been applied in other AI (machine learning) contexts where the available training datasets underrepresent a group (Chawla et al., 2002). However, in the current context, there is a risk that this approach might cause the AI to overfit its predictions to the properties of the specific successful women in this set, because the set would not reflect the diversity of properties that would be present if other women were included. Another approach is to create synthetic data to supplement the category that is underrepresented. For example, when facial detection algorithms exhibited lower accuracy in classifying faces from persons of color because images in the dataset were mostly of Caucasian males, a solution involved creating more synthetic faces of people from other races to include in the training data for the system (Wachter-Boettcher, 2017).

Bias in Measures of Predicted Performance. Even among the individuals included in datasets, the performance or "fit" measures used to rate or rank these individuals might be biased indicators (Tambe et al., 2019). Evaluations by humans or human-designed assessment tests are vulnerable to individual and systemic bias. For example, Elvira and Town (2001) found that when actual performance was equal, black employees tended to receive lower ratings than white employees, and Marlowe et al. (1996) found that gender and attractiveness can affect the ratings and rank-ordering of candidates. As some men tend to consider women less physically attractive as they age (Mathes et al., 2010), this latter bias might especially disadvantage more mature female candidates. The data set or AI system might not have explicitly coded for the race, gender, or attractiveness of candidates, but in some cases the system can learn to associate a low rating to with a related available input, e.g., year of college graduation is related to age, and which college the candidate attended can be related to racial or gender demographics. More generally, other available attributes that actually predict performance may become undervalued by the system because they are associated with individuals in the training set whose ratings were inappropriately low due to bias. If the training set provided to the AI includes biased performance measures, the AI will emulate this bias and not make optimal selections. This discussion emphasizes that the selection behavior of algorithms is exceedingly sensitive to the choice of measure of employee performance and/or fit that is used to train them (Kleinberg et al., 2018b).

To address this source of bias, it has been suggested that potential training datasets (performance measures) could be possibly checked for biases via statistical analysis prior to being used to train AIs (Miller et al., 2018). For example, one could examine and compare the distributions of performance/fit ratings in the data set by sex, race, religion, etc. The question becomes how to use this information. If one were to eschew using all datasets that manifested some bias, it is likely one would reject all available datasets. Alternatively, one might try to adjust the dataset to "correct" the bias, but the most appropriate method to do so might be unclear, and might have unforeseen consequences when the set is used to train an AI (Hajian et al., 2016). As an example, one might pre-process that data to adjust the performance ratings of minorities and women in the data set so that the distributions are more similar to the distribution for white males (if they have systematically higher ratings). Another possibility is to check for and try to resolve bias once the candidate ratings from the AI are received. If equal opportunity is a company policy, one might re-rank the recommendation list in accordance with this priority. Finally, as mentioned earlier, one might apply such bias detection techniques not just to the training data inputs of a system but also to the output recommendations of a system (e.g., Facebook's Fairness Flow; Meyer, 2018).

Bias in Candidate Properties Used to Predict Performance. One might wonder why we could not effectively resolve dataset bias if we could just make the AI blind to the sex, race, religion, etc. of candidates in the training set (and the new candidates we want the AI to categorize for us). Thus, all candidates (past and future) would effectively be viewed by the AI to have the same sex, race, religion, etc. There are two possible problems with this approach. First, it turns out that "blinding" the system to such information can be easier said than done. Second, if a goal might be to increase equity in the selection process (e.g., reduce the discrepancy between the proportion of white and black candidates selected), it might actually be necessary/appropriate to include racial information in the system inputs (Kleinberg et al., 2018a). We will talk about these two issues in turn.

Consider the case of that Amazon AI system that systematically issued lower ratings to female candidates. Even if the system was not given an explicit input specifying a candidate's sex, the system was able to effectively "infer" this information from indirect evidence in the resume, such as the fact that the candidate listed a college that happened to be an all-women's college, or mentioned membership in a women's team or club (Dastin, 2018). Thus, even if protected features are not explicitly included in the inputs, there remains what is known as a *reconstruction problem*, whereby the system can effectively glean (i.e., reconstruct) protected features from other inputs (Kleinberg et al., 2018a). We suggest that the term *reconstruction* is, however, somewhat misleading. The system was not aware that the candidate was female *per se*. Rather the system detected that some of the attributes of the candidate did not optimally match with typical attributes of high performing employees from the training database. As the training database was comprised predominantly of men, it is unsurprising that high performers in that set were not graduates of all-women's colleges, nor were they members of women's clubs or teams. Even if applicants or AI developers delete such specific "tells" we cannot be confident that other information available to the system might continue to contribute to an implicit bias because the candidate has attributes that do not most closely match the narrow (e.g., white and/or male) pool of successful employees in the database.

Recall that AI HR systems might have access to information beyond what candidates explicitly provide for job application purposes (e.g., résumés and interview videos). For example, some recruiting systems like TalentBin use social media data to match potential candidates to jobs. AI systems can use information available about us in digital records of our behavior like social media to infer private and protected traits about us including (but not limited to): sex, race, sexual orientation, and religious and political views (Kosinski et al., 2013 cited in Raub, 2018). Worse still, employers can (or at least could, according to these sources) post job ads on Facebook and explicitly target them to certain sets of individuals based on demographic information known to Facebook such as age, religion, relationship status, political affiliation, and ethnic affinity (Hao, 2019; Kim, 2020). Even if ads are not explicitly set up to target certain groups, the Facebook AI algorithm that matches ads to individuals still exhibits bias—for example, jobs for nurses and secretaries tend to be shown to more female users and jobs for taxi drivers and janitors tend to be shown for a larger fraction of minority users (Hao, 2019). This bias arises because the AI was trained on historical Facebook data where individuals from those groups were more likely to click on ads for those types of jobs. However, the AI matching algorithm then reinforces and perpetuates such trends. Facebook was reportedly testing a tool called Fairness Flow to spot biases (e.g., age, race, gender) in machine-learning algorithms in general, including their algorithm for matching job ads with job seekers (Gershgorn, 2018; Meyer, 2018).

3.3 Bias in Interpretation and Use of AI Recommendations

One concern about the use of AI systems for such impactful decisions is that the AI will be trusted implicitly, and that companies might fail to assess and therefore recognize bias in the system. This tendency to have too much faith in automated systems is called automation bias (Robineete et al., 2016 cited in Gebru, 2019). Presently, however, those that develop and use such AI systems make clear that AIs do not make the final call in HR decisions (e.g., Captain, 2016). Rather, a human or team of humans typically makes a selection from among a set of candidates who received top rankings from the AI system.

Such "human-in-the-loop" decision making might be comforting to those who find such computational systems dehumanizing (Eubanks, 2018), and those who believe that candidates will more readily accept the involvement of an AI in a selection process if they know that a human has a final say over the outcome (e.g., Tambe et al., 2019). However, given that bias in AI systems is typically introduced by training them on prior human decision patterns, it might seem a bit surprising that human involvement would be so reassuring. Nonetheless, if end-stage human involvement is reassuring, it might benefit the company by reducing litigation and accusations of discriminatory hiring practices. However, there is reason to believe that end-stage human involvement could sometimes actually introduce bias.

Interviews are a common means to select a candidate from a short list (e.g., the AI's top ranked candidates), however, the reliability of interviews is limited (Van Iddekinge et al., 2006). The evaluation of a candidate can vary significantly depending on the interviewer. This is especially true for unstructured interviews which might not involve a consistent evaluation rubric or set of questions across candidates. Conway et al. (1995) found that while structured interviews resulted in a reasonable degree of consistency across interviewers (inter-rater reliability of 0.67), unstructured interviews were associated with much lower reliability (0.34). That said, even structured interviews are not free from bias. For example, even when interviews were structured and included quantitative measures, one study found that UK military interviewers were inclined to provide more positive assessments of individuals that shared similar features to their own (e.g., having attended the same type of school; Salaman & Thompson, 1978). Despite the fact that quantitative evidence was available, the assessors either emphasized or deemphasized characteristics of candidates to rationalize their ratings, e.g., "lacks natural authority", their experience is too "narrow", or "he certainly fit in". Other experimental studies have suggested that job-irrelevant features of a candidate (e.g., weight) can affect judgments of merit (Angerstrom & Rooth, 2011). Thus, there is the possibility that human involvement in the final stages of the selection process has the potential to introduce rather than reduce bias.

4 Summary and Conclusions

AI has the potential to make some HR processes more efficient and to reduce bias. To capitalize on this potential, we must be cognizant to avoid and/or mitigate biases that might be introduced by human influences such as: designers and design choices; training data based on past hiring patterns by humans; and human involvement in selecting a final candidate from the system's recommendations. To reduce bias in design, AI developers should ideally be more demographically diverse, and either have some training to detect potential bias that might arise from design choices, or as we suggest, be teamed with an HR professional trained in such detection.

Training data might involve biased performance measures and be demographically incomplete or imbalanced. Statistical and machine learning methods can play a key role in identifying such issues with the training data, and with system recommendation patterns that mirror past human practices. Methods such as oversampling and the inclusion of synthetic data can contribute to mitigating some training data imbalances. In terms of the system's recommendation patterns, although explainable AI research has not reached the stage which renders the system's decision criteria transparent, the use of fairness metrics (e.g., group fairness, predictive equality) can help assess whether it is behaving in accordance with legal and ethical expectations (i.e., quality assurance). Such assessment is key before prototype systems are put into practice. Furthermore, note that such quality assurance checks are much easier to do on an AI system than a human hiring manager. The AI can quickly process a large test set of hypothetical candidates to reveal trends, and, unlike a human, the AI will not alter its behavior in an assessment context to "game" the test. For human hiring managers, bias might be reduced by ensuring they are unaware of a candidate's race and sex. Counter-intuitively, to reduce bias and meet fairness metrics, some AI systems might require protected factors (e.g., race, sex) to be available as inputs to enable system designs that explicitly attempt to counteract bias present in training datasets.

Overall, AI systems hold tremendous promise because they can contribute in at least three ways to reducing bias in HR. Scrutiny arising from the introduction of AI into HR has already served to expose bias in human hiring practices. AI and machine learning systems can detect bias in training datasets and recommendation patterns and AI selection systems can be designed that need not emulate the often biased human selection practices. However, further consideration is required in order to specify how the social and ethical issues that are generally relevant to AI systems can be addressed in the context of HR.

References

Abdi, H. (2007). Signal detection theory (SDT). In B. McGaw, P. L. Peterson, & E. Baker (eds), *Encyclopedia of Measurement and Statistics* (3rd Ed., 886–889). Elsevier.

Adadi, A., & Berrada, M. (2018). Peeking inside the black-box: A survey on Explainable Artificial Intelligence (XAI). *IEEE Access*, 17(6), 52138–52160.

Agerström, J., & Rooth, D. O. (2011). The role of automatic obesity stereotypes in real hiring discrimination. *Journal of Applied Psychology*, 96(4), 790–805.

Aspan, M. (2020, January). This tech giant says A.I. has already helped it save $1 billion. *Fortune*. https://fortune.com/2020/01/24/ai-ibm-human-resources/.

Barnard, D. (2019, August) Examples of how AI is transforming learning and development. *Virtual Speech*. https://virtualspeech.com/blog/ai-ml-learning-development.

Bertrand, M., & Mullainathan, S. (2004). Are Emily and Greg more employable than Lakisha and Jamal? A field experiment on labor market discrimination. *American Economic Review*, 94(4), 991–1013.

Captain, S. (2016, May 18). Can artificial intelligence make hiring less biased? *Fast Company.* https://www.fastcompany.com/3059773/we-tested-artificial-intelligence-platformsto-see-if-theyre-really-less-.
Cath, C. (2018). Governing artificial intelligence: Ethical, legal and technical opportunities and challenges. *Philosophical Transactions A, 376*, 20180080. http://dx.doi.org/10.1098/rsta.2018.0080.
Chawla, N. V., Bowyer, K. W., Hall, L. O., & Kegelmeyer, W. P. (2002). Smote: Synthetic minority over-sampling technique. *Journal of Artificial Intelligence Research, 16,* 321–357.
Conway, J. M., Jako, R. A., & Goodman, D. F. (1995). A Meta-analysis of interrater and internal consistency reliability of selection interviews. *Journal of Applied Psychology, 80*(5), 565–579.
Core, M. G., Lane, H. C., Van Lent, M., Gomboc, D., Solomon, S., & Rosenberg, M. (2006, July). Building explainable artificial intelligence systems. In *AAAI* (pp. 1766–1773).
Dastin, J. (2018). Amazon scraps secret AI recruiting tool that showed bias against women. *Reuters.* https://www.reuters.com/article/us-amazon-com-jobs-automation-insight/amazon-scraps-secret-ai-recruiting-tool-that-showed-bias-against-women-idUSKCN1MK08G.
Elvira, M., & Town, R. (2001). The Effects of race and worker productivity on performance evaluations. *Industrial Relations: A Journal of Economy and Society, 40*(4), 571–590.
Eubanks, V. (2018). *Automating inequality: How high-tech tools profile, police, and punish the poor.* St. Martin's Press.
Gebru, T. (2019). *Oxford handbook on AI ethics book chapter on race and gender.* arXiv:1908.06165.
Gee, M. (2018). Why aren't Black employees getting more White-collar jobs. *Harvard Business Review.* https://hbr.org/2018/02/why-arent-black-employees-getting-more-white-collar-jobs.
Gershgorn, D. (2018, May). Facebook says it has a tool to detect bias in its artificial intelligence. *Quartz.* https://qz.com/1268520/facebook-says-it-has-a-tool-to-detect-bias-in-its-artificial-intelligence/.
Greene, D., Hoffmann, A. L., & Stark, L. (2019, January). Better, nicer, clearer, fairer: A critical assessment of the movement for ethical artificial intelligence and machine learning. In *Proceedings of the 52nd Hawaii International Conference on System Sciences* (pp. 2122–2131). Hawaii, USA.
Hajian, S., Bonchi, F., & Castillo, C. (2016, August). Algorithmic bias: From discrimination discovery to fairness-aware data mining. In *Proceedings of the 22nd ACM SIGKDD International Conference on Knowledge Discovery and Data Mining* (pp. 2125–2126). San Francisco, CA.
Hao, K. (2019, April). Facebook's ad-serving algorithm discriminates by gender and race. *MIT Technology Review.* https://www.technologyreview.com/2019/04/05/1175/facebook-algorithm-discriminates-ai-bias/.
Hardt, M., Price, E., & Srebro, N. (2016). Equality of opportunity in supervised learning. In *Advances in Neural Information Processing Systems* (pp. 3315–3323).
Jobin, A., Ienca, M., & Vayena, E. (2019). The global landscape of AI ethics guidelines. *Nature Machine Intelligence, 1*(9), 389–399.

Kim, L. (2020, April). How to use Facebook ads to recruit top talent. *The WordStream Blog.* https://www.wordstream.com/blog/ws/2016/08/23/facebook-recruiting.

Kleinberg, J., Ludwig, J., Mullainathan, S., & Rambachan, A. (2018a, May). Algorithmic fairness. In *AEA Papers and Proceedings* (Vol. 108, pp. 22–27). https://doi.org/10.1257/pandp.20181018.

Kleinberg, J., Ludwig, Mullainathan, S., & Sunstein, C. (2018b). Discrimination in the age of algorithms. *Journal of Legal Analysis, 10,* 113–174. https://doi.org/10.1093/jla/laz001.

Knight, W. (2017, April 11). The dark secret at the heart of AI: No one really knows how the most advanced algorithms do what they do. That could be a problem. *MIT Technology Review.* https://www.technologyreview.com/s/604087/the-dark-secret-at-the-heart-of-ai/.

Kosinski, M., Stillwell, D., & Graepel, T. (2013). Private traits and attributes are predictable from digital records of human behavior. *Proceedings of the National Academy of Sciences, 110*(15), 5802–5805.

Kuhn, P., & Shen, K. (2013). Gender discrimination in job ads: Evidence from china. *The Quarterly Journal of Economics, 128*(1), 287–336.

LinkedIn (2018a). *Global recruiting trends.* https://news.linkedin.com/2018/1/global-recruiting-trends-2018.

LinkedIn (2018b). *The rise of HR analytics.* https://business.linkedin.com/content/dam/me/business/en-us/talent-solutions/talent-intelligence/workforce/pdfs/Final_v2_NAMER_Rise-of-Analytics-Report.pdf.

Marlowe, C. M., Schneider, S. L., & Nelson, C. E. (1996). Gender and attractiveness biases in hiring decisions: Are more experienced managers less biased? *Journal of Applied Psychology, 81*(1), 11.

Mathes, E., Brennan, S., Haugen, P. & Rice, H. (1985). Ratings of physical attractiveness as a function of age. *The Journal of Social Psychology, 125*(2), 157–168.

Meyer, D. (2018, Oct.). Amazon reportedly killed an AI recruitment system because it couldn't stop the tool from discriminating against women. *Fortune.* https://fortune.com/2018/10/10/amazon-ai-recruitment-bias-women-sexist/.

Miller, F. A., Katz, J. H., & Gans, R. (2018). The OD imperative to add inclusion to the algorithms of artificial intelligence. *OD Practitioner, 50*(1), 6–12.

Mundy, L. (2017, April). Why is Silicon Valley so awful to women? *The Atlantic.* https://www.theatlantic.com/magazine/archive/2017/04/why-is-silicon-valley-so-awful-to-women/517788/.

Munoz, C., Smith, M. & Patil, D.J. (2016, May). Big Data: A report on algorithmic systems, opportunity, and civil rights. *Executive Office of the President.* https://obamawhitehouse.archives.gov/sites/default/files/microsites/ostp/2016_0504_data_discrimination.pdf.

Ras, G., van Gerven, & M., Haselager, P. (2018). Explanation methods in deep learning: Users, values, concerns, and challenges. In H. Escalante (ed.), *Explainable and interpretable models in computer vision and machine learning* (pp. 19–36). Springer.

Raub, M. (2018). Bots, bias and big data: Artificial intelligence, algorithmic bias and disparate impact liability in hiring practices. *Arkansas Law Review, 71,* 529.

Ribeiro, M., Singh, S., & Guestrin, C. (2016). "Why should I trust you?" Explaining the predictions of any classifier. In *ACM SIGKDD Conference on Knowledge Discovery and Data Mining (KDD)*. (pp. 1135–1144). San Francisco, CA.

Robinette, P., Li, W., Allen, R., Howard, A. & Wagner, A. (2016). Overtrust of robots inemergency evacuation scenarios. In *Proceedings of the Eleventh ACM/IEEE International Conference on Human Robot Interaction* (pp. 101–108). IEEE Press.

Rosenbaum, E. (2019, April). IBM artificial intelligence can predict with 95% accuracy which workers are about to quit their jobs. *CNBC*. https://www.cnbc.com/2019/04/03/ibm-ai-can-predict-with-95-percent-accuracy-which-employees-will-quit.html.

Salaman, G., & Thompson, K. (1978). Class culture and the persistence of an elite: The case of army officer selection. *The Sociological Review*, 26(2), 283–304.

Samek, W., & Müller, K. (2019). Towards explainable artificial intelligence. In W. Samek, G. Motavon, A. Vedaldi, L. K. Hansen, & K. Muller (eds), *Explainable AI: Interpreting, explaining and visualizing deep learning* (pp. 5–22). Springer.

Schneider, M. (2017, July). Google gets 2 million applications a year: To have a shot, your resume must pass the '6-second test'. *Inc*. https://www.inc.com/michael-schneider/its-harder-to-get-into-google-than-harvard.html.

Silberg J., & Manyika, J. (2019, June). Notes from the AI frontier: Tackling bias in AI (and in humans). *McKinsey Global Institute*. https://www.mckinsey.com/featured-insights/artificial-intelligence/tackling-bias-in-artificial-intelligence-and-in-humans.

Simon, H. A. (1997). *Administrative behavior*. Simon and Schuster.

Staats, C., Capatosto, K., Wright, R. A., & Jackson, V. W. (2016). *State of the science: Implicit bias review*. Kirwan Institute for the Study of Race and Ethnicity.

Su, G. (2018). Unemployment in the AI age. *AI Matters*, 3(4), 35–43.

Tambe, P., Cappelli, P., & Yakubovich, V. (2019). Artificial intelligence in human resources management: Challenges and a path forward. *California Management Review*, 61(4), 15–42.

Thomson, R., & Schoenherr, J. R. (2020, July). Knowledge-to-information translation training (KITT): An adaptive approach to explainable artificial intelligence. In *International Conference on Human-Computer Interaction* (pp. 187–204). Springer.

Thomson R., & Schoenherr J. R. (2020). Knowledge-to-information translation training (KITT): An adaptive approach to explainable artificial intelligence. In R. Sottilare & J. Schwarz (eds.) *Proceedings from Adaptive Instructional Systems, Lecture Notes in Computer Science* (Vol. 12214). Springer. https://doi.org/10.1007/978-3-030-50788-6_14.

Thorngate, W. (1988). On the evolution of adjudicated contests and the principle of invidious. *Journal of Behavioral Decision Making*, 1, 5–15.

Thorngate, W., & Carroll, B. (1990). Tests versus contests: A theory of adjudication. In W. Baker (ed.), *Recent trends in theoretical psychology* (pp. 431–438). Springer.

Tiku, N. (2018, June). Google's diversity stats are still very dismal. *Wired*. https://www.wired.com/story/googles-employee-diversity-numbers-havent-really-improved/.

Toossi, M. (2002). A century of change: The U.S. labor force, 1950–2050. *Monthly Labor Review, 125*(5), 15–28.

Van Iddekinge, C. H., Sager, C. E., Burnfield, J. L., & Heffner, T. S. (2006). The variability of criterion-related validity estimates among interviewers and interview panels. *International Journal of Selection and Assessment, 14*(3), 193–205.

Verma, S., & Rubin, J. (2018, May). Fairness definitions explained. In *2018 IEEE/ACM International Workshop on Software Fairness (FairWare)* (pp. 1–7). IEEE.

Wachter-Boettcher, S. (2017). *Technically wrong: Sexist apps, biased algorithms, and other threats of toxic tech*. WW Norton & Company.

Chapter 10

On the Shoulders of Giants
How the Social Sciences Can Help AI Navigate Its Ethical Dilemmas

Arthur C. Graesser and John Sabatini

1 Introduction

The primary goal of this *Frontlines of AI Ethics* book is to explore how the field of psychology and other social sciences can help the field of Artificial Intelligence (AI) navigate complex questions about the ethics of the systems they develop. The power and utility of some AI systems are undeniable. There are AI systems that drive vehicles, perform surgical procedures, grade essays, and hold spoken conversations in natural language. However, citizens are raising dozens of questions about the ethics of these AI systems. Concerns are typically launched when an AI system makes an error: an accident involving a self-driving car, an error in a surgical procedure, a low essay grade on a high stakes test, and a bot that allegedly leads to a financial loss or even a suicide. Errors sometimes occur in AI systems, even when there are far more errors by humans. Comparative error rates can be made between AI systems and humans, of course, and everyone agrees that the errors of AI systems should be reduced as far as possible. However, the criteria and biases that underlie error-prone actions must also align with ethical standards. This book addresses such standards from the standpoint of psychology as well as other social sciences. Perhaps AI can adopt some of them.

It is important to acknowledge that ethical policy and guidelines in both AI and the social sciences are in a state of flux, so the details expressed in this book are frozen at a point in history. At the time this chapter was written, all psychological studies are required to pass an Institutional Review Board (IRB) on the ethics of the research, which follow standards of practice (American Educational Research Association, American Psychological Association, & National Council on Measurement in Education, 1999) and methodology (Gersten & Hitchcock, 2009). The IRB serves a gatekeeper function regarding social science research. The board is composed of five individuals in academia and the public who comment on the research and vote whether it is approved. Every researcher must undergo training and certification in ethics of conducting research. In contrast, private industries (large and small) that do not seek federal funding do not have the constraint of an IRB; they often express that they have an internal committee that reviews ethical procedures but are not always transparent on what that is. Therefore, at the onset, there is a political and legal bias that requires stringent criteria for university research but loose or nonexistent criteria for industry, especially in a new frontier of investigation like AI. How many of the surveys on our smart phones had to pass an IRB panel on ethical concerns? The answer is zero unless it involves university and grant-funded research.

Many AI researchers are concerned about ethical policies and guidelines so one first place to look is the social sciences, especially psychological investigation that directly considers the potential human cost of novel technologies. The chapters in this volume offer some insights. To summarize, the chapters consider AI from the perspectives of philosophy and history (Chapter 1), data science (Chapter 2), potential biases in linguistic analysis of political discourse (Chapter 3), intelligent tutoring systems (Chapter 4), deep learning systems (Chapter 5), delivery of educational content (Chapter 6), the training mission of the military (Chapter 7), cybersecurity (Chapter 8), and human resources (Chapter 9). The authors vary in their positions on ethical applications and implications of AI. They identify risks and areas of concern, but most of these can be managed with guidelines and insights, especially from the cognitive and social sciences. The authors of some chapters identify invisible risks and recommend more stringent safeguards to AI development and deployment.

In this chapter, we reflect on how the ideas in this book convey the lessons of social science research into new questions arising in AI. These reflections are organized around five questions. In what ways are humans and AI systems rational? What forms of bias exist in humans and AI systems? How can one handle the indeterminacy of meaning among humans and in AI systems? How can computer agents be compared with human conversation partners? How can humans and AI systems promote beneficence and dignity for humans? We conclude that there are no perfect answers to any of these questions, but rather there are tradeoffs that need to be scrutinized by humans and quantitative computational systems in a hybrid evaluation system.

2 In What Ways Are Humans and AI Systems Rational?

Several chapters in this volume (Ulgen; Ness; Pyke, Schoenherr, & Thomson) commented on the extent to which humans versus AI systems exhibit rational reasoning. The symbolic and statistical algorithms in AI systems are systematic, but there is always the worry of whether the algorithms are relevant, complete, minimize bias, and optimize human beneficence. This is an important issue to address because the contributors had some disagreement on the nature and quality of rationality exhibited by humans and AI. This debate, of course, bolsters the proposal to have a hybrid process that includes collaborative interaction between humans trained in ethical principles and output of computational systems.

We know from the last 50–200 years that individual humans do not routinely exhibit rational reasoning and behavior. Instead, they predominantly activate simple heuristics (Kahneman, 2011) and ideas that resonate with unconscious funds of knowledge that reflect a person's years of experience in particular sociocultural contexts (Gladwell, 2019). Individuals are highly influenced by others in groups with which they affiliate, a form of tribalism. Following that understanding, the psychological and historical evidence is not strongly in favor of humans (individually or collectively) arriving at rational resolutions of ethical dilemmas. We should at least entertain the proposition that assistance from AI might be beneficial. A central argument in this chapter is that there are roles for AI systems that could subserve human decision making in support of ethical policies and outcomes. It is worthwhile to explore where AI may prove beneficial.

Mundane plausible reasoning in humans deviates radically from formal reasoning. Humans do well on *modus ponens* (If X, then Y; X; therefore Y), consistently fail on *modus tollens* (If X, then Y; not Y; therefore not X), and often embrace the abductive reasoning that has an illegitimate formal logical foundation (Rips, 1994). For example, person (A) approaches another person (B) and dramatically expresses "Stop or I'll shoot." Person B stops, but person A shoots B anyway. Person A's shooting B is logically legitimate. However, humans get upset because they are trapped by biconditional symmetry (both $X \square Y$ and $Y \square X$ are correct) and have no routine capacity for implementing *modus tollens*, as well as many other rules of formal reasoning (e.g., De Morgan's rule). *Modus ponens* can be handled by humans because it is simply a pattern match or resonance process between the stimulus input (e.g., the given information, X) and the knowledge repository ($X \square Y$ *and X, therefore Y*). They have trouble with negations that are required by modus tollens ($X \square Y$ and $\square Y$, *therefore* $\square X$). They have difficulty with implicit hypothetical worlds and updating the changes in hypothetical worlds (Johnson-Laird, 2010). In contrast, computers have the capacity of faithfully implementing many rules in theorem provers, including negation and legal hypothetical chains of consequences (Newell & Simon, 1972). Logical reasoning in humans is not an inherent habit of mind but can be achieved by effortful use of external artifacts that range from truth tables, logical path trajectories, proofs, and AI simulations of predictions according to multiple possible worlds. A single human cannot do it alone. Thus, AI augmentation is likely able to support more rational decision-making whenever systematic, logical analyses are considered important to the decision process.

There are other examples that support the point that humans are not routinely rational. Statistical reasoning also diverges between formal systems and humans, who are prone to have, for example, base-rate and hindsight biases (Kahneman et al., 1982). Humans are highly influenced by anecdotes about unusually interesting single cases, even though single cases have no relevance in statistical reasoning about samples of cases. In the area of attitude change, humans are primarily persuaded by peripheral features (frequency, salience, emotions, anecdotes, stories) rather than coherent arguments with evidence (Petty & Cacioppo, 1986). Such facts, of course, have relevance to what humans versus AI/robot systems can accomplish. Humans alone are not the gold standard. There are many cases where statistical probability and learning should be considered in arguing for or against a pathway or policy, whether ethical or pragmatic. Humans may overrule such arguments to go against the odds, but they should not falsely believe that the odds are in favor of their choices. Humans must realize they fall prey to flawed human cognitive heuristics, especially when psychology has identified and quantified when those flaws are counter to the evidence.

Perhaps a highly trained and educated group of humans can do a better job in providing rational analyses whereas individuals fail. Unfortunately, research on collaborative decision making and collaborative problem-solving reveal that small groups are highly influenced by conformity and "groupthink" (Graesser et al., 2018; Janis, 1982). Large groups are not necessarily an added benefit in attempts to achieve logical rational solutions. Large groups are too often influenced by politics rather than rational solutions, as is apparent in the 50–50 split of the US Senate that follows party lines, arguably without any principled rational foundation beyond the us versus them mentality endemic to tribal civilizations. On the optimistic side, Pinker (2018) has reminded us that slow progress on rational solutions to societies' problems have been made cumulatively over decades or centuries when large groups follow principles of rationality. Again, human judges operate best when the evidence they have access to is as complete and sound as possible. This is the grist of sound decision-making by groups or individuals.

It is important to be clear on the status of rationality in AI. Modern AI systems are not confined to rigid, brittle, rational, deterministic systems, as some critics contend. For over three decades, the new AI (Franklin, 1995) has explored neural networks, deep learning, dynamical systems, multiagent systems, fuzzy logic, and complex systems that differ dramatically from the "old" AI. These intelligent systems are complex—far more complex than a simple rule-based deterministic mechanism that is relatively easy to follow or reconstruct. They can draw probabilistic conclusions, with odds of different possible outcomes estimated. Indeed, they are so complex that it is difficult (but not always impossible) to explain how the system arrived at a major decision. One of the current trends in AI is to develop compelling explanations of solutions to problems so that black box results are scrutable. AI results or probability estimates can be quantified using sophisticated statistical inference, error bands, and effect sizes, as in all modern social science research outcomes. Thus, AI need not be any more ungoverned than the human decision-making process, which itself is often described as intuitive or unconsciously derived.

Nearly three decades ago, there was a request to review a book written by Alexander Silverman on AI and legal reasoning (Graesser, 1995). The book (Silverman, 1993) compellingly made the argument that simple brittle rules encounter problems in legal reasoning because of all sorts of exceptions that cannot be envisaged ahead of time. A simple rule like "vehicles are not allowed in the park" sounds initially wise and high in beneficence, but exceptions quickly present themselves: toy cars operated by children, floats in a parade on the fourth of July, a fire truck to put out a fire, a citizen driving her truck to pull off a log that fell on a person's leg. There are a potentially infinite number of exceptions that disqualify a simple rule, far too many to keep track of in a legal system or even a computer program. One of the humorous examples in AI is the "potato in the tailpipe problem." A car will not start if a potato is placed in the tailpipe. However, many/most AI systems would not check for that situation when diagnosing a vehicle that will not start.

Silverman explored the prospects of a neural network (NN) replacing a brittle rule-based system in legal reasoning. Dozens/hundreds of input features would be taken into consideration and collectively produce an output of legal relevance (e.g., infraction versus not, guilty versus not). A neural network could be quite accurate if it is based on a large enough corpus, which is the hallmark of modern Deep Learning efforts (Ahad et al., 2018). Unfortunately, there are problems with this approach in the arena of law. How could an attorney defend the reasoning of a NN in a courtroom? Law has a case-based foundation, not a probabilistic statistical foundation. Attorneys are not routinely trained to make decisions on quantitative tradeoffs and cost-benefit analyses. It also would take much too much time to sort out whether the input features of a NN are legitimate for a specific case under consideration. Further, a biased sample of cases could adversely influence the training of such a NN, as discussed in the next section. Modern AI systems operate across computationally complex corpora of instances, searching for patterns and exceptions. Imagine a system that generated a list of every published court case exception to a specific law, and provided that list to prosecution, defense, and judge; this could include the odds ratio of how likely an exception was found to an application of the law. As long as we avoid a total subjugation of human ethical decision making in favor of AI-derived algorithmic conclusions, we have the opportunity of humans to aspire to a more rationally informed set of ethical decisions.

The chapters in this book illustrate many of the complexities of establishing an ethical foundation of AI. This section attempts to clarify some limits of rationality in both humans and AI. Individual humans, groups of humans, and AI systems all face obstacles to rationality, however that may be defined. There presumably is value in a hybrid approach that considers analyses from humans with different perspectives and AI systems with different models.

3 What Forms of Bias Exist in Humans and AI Systems?

Bias in humans has been extensively studied for several decades in the psychology of decision making (Kahneman et al., 1982; 2021) and social cognition (Fiske, 1998). It is well established that humans are substantially influenced by stereotypes about race, ethnicity, gender, age, country of origin, the list goes on. It is difficult for humans to emancipate themselves from these stereotypes because they are built from decades of experience unconsciously and occasionally consciously. Data show that people believe that female librarians drink wine and female truck drivers drink beer, but most of the respondents have never met a female librarian or truck driver. Interestingly, it could be argued that AI systems are victims of "technology stereotyped threat" that typecasts these systems as deterministic, rigid, mechanical, uncaring, narrow, the list goes on. However, AI systems have been developed for decades that challenge these attributes, including our research that tracks students' engagement and emotions in a system called AutoTutor that helps students learn with conversational agents (Chen et al., 2021; D'Mello & Graesser, 2012).

Many of the chapters in this volume address bias (D. DeFalco; Pyke, Schoenherr, & Thomson; Schoenherr & Thomson; Windsor). The foundations of bias are multifaceted: bias in the sampling of observations, bias of people selected, media bias, historical-temporal bias, various types of semantic bias, technological bias, etc. Potentially an infinite number of bias categories and dimensions stand ready for exploration. How are researchers able to handle this landscape of sources of bias? All of the chapters in this volume addressed the problem of bias, but they were not in agreement on the relative biases in humans versus AI systems. Some of the contributors may not have been aware of the major advances in techniques to minimize bias in AI systems, and perhaps under-estimated the profound biases in humans. For example, most machine analyses in AI sample observations systematically in a manner that equilibrates different populations of learners (placing them on an even playing field) rather than having solutions that simply reflect selection biases of the most frequent populations (Baker, 2020).

Unfortunately, individuals and groups of humans are ill-equipped to handle such multifaceted variability. There may be ways for AI and other quantitative analyses to help navigate this rugged terrain. A *differential item* analysis is routinely conducted in psychometric research on assessments that assess whether populations are unfairly scored on particular items and tests (Reynolds et al., 2021). Clustering analysis in AI, data mining, and psychometrics can identify clusters of individuals who exhibit problems on a test. In fact, much of the methodological and statistical machinery of the social sciences is designed to detect and protect against bias. The concept of representative sampling is a safeguard against over-generalizing a result based on a bias in the sample. The classic work of Campbell and Cook (1979) on threats to validity in psychological experimentation or Messick's (1995) work on the validity of psychological assessment stand out as cornerstones in a long line of research in how to ethically guide the process and product of social science research toward valid and trustworthy outcomes. Once again, such data need to be considered in a sophisticated and humane analysis of bias. It is not sufficient to rely only on a small committee of humans (e.g., an IRB panel) to provide significantly disparate points of view, based on their idiosyncratic life histories. Fortunately, we have sophisticated tools in the researcher's kit to evaluate and quantify the effectiveness, validity, and fairness of the AI systems we might choose to deploy (e.g., American Educational Research Association, American Psychological Association, & National Council on Measurement in Education, 1999; Gersten & Hitchcock, 2009; What Works Clearinghouse, 2012).

Unlike human decision making, the field of AI draws from this research history to take systematic actions to minimize bias. There is careful consideration of the properties of the data in applications of machine learning. Every computer scientist is aware of the expression "garbage in, garbage out (GIGO)" that succinctly conveys the point that flawed or nonsense input data produce nonsense output or "garbage." Similarly, expert researchers in data mining and machine learning are sensitive to the notion of "data bias in, bias out" (Baker, 2020). Consequently, there is a process of carefully examining the distribution of data to see whether different populations are adequately represented. They make statistical adjustments when data samples deviate from population distributions. They assess whether data samples are sufficiently large to offer conclusions and compare training samples with test samples to assess the generalizability of findings. When a critic claims that an AI solution is biased on feature F (e.g., gender, race, socioeconomic status), there is a systematic quantitative procedure to evaluate the accusation. Consequently, there is accountability in analyses of bias. In contrast, humans lack a comparable degree of accountability, if any accountability at all other than compatibility with dubious biased stereotypes. That is why a hybrid approach to ethical decision making is prudent: humans using AI and quantitative data.

One limitation of this hybrid approach is that the humans will not trust the output from AI systems. Humans tend to insist their biases are true in the face of counter-evidence (Rapp & Braasch, 2015), which would include mistrusting the AI-generated evidence. The hope is that humans will eventually converge on wise decisions with a hybrid model that uncovers advantages and limitations of both humans and computational solutions.

4 How Can One Handle the Indeterminacy of Meaning among Humans and in AI Systems?

Several AI applications analyze and generate natural language, as in the case of automated essay grading, automated event generation in news reports, language translation, speech recognition, and information extraction from conversation. Unfortunately, AI analyses of natural language meaning are imperfect. Similarly, humans often disagree on the meaning of words, sentences, and discourse, so there is a big challenge in handling the indeterminacy of meaning. Much of the language that humans generate is ambiguous, vague, underspecified, fragmentary, elliptical, and contradictory (sometimes intentionally). Comparisons between cultures and languages, of course, contribute to the problem of indeterminacy, adding to its ubiquity. Some of the chapters in this volume (Ness, Burrell, & Frey; Hampton, Morrison, & Morgan; Windsor) raised worries about the indeterminacy of meaning. They also disagreed on the quality of AI systems in conducting analyses of meaning.

Empirical analyses of meaning involve a group of humans annotating samples of language/discourse segments (words, sentences, texts) on particular language or discourse features, such as a category of speech act (e.g., assertion, question, expressive evaluation), syntactic complexity, topic relevance, emotion sentiment, and so forth. Alternatively, pairs of language segments are judged on similarity of meaning. It is important to emphasize that human judges, even trained expert human judges, do not show high agreement on many categories of judgments. As an example, we have conducted research in computational linguistics and intelligent tutoring systems with natural language in several projects. We have reported interrater agreement scores for syntactic complexity at a modest 0.40 for experts in linguistics (Graesser et al., 2000), 0.69 for semantic similarity judgments on statements in scientific reasoning (Cai et al., 2011), and 0.30–0.40 on whether a facial expression manifests a particular emotion during learning (D'Mello & Graesser, 2012). These scores are largely unimpressive, and certainly far from perfect. Equally interesting are the empirical reports that comparisons between computer-generated judgments and human judgments are nearly as high as pairs of humans (Yan et al., 2020). Automated essay grading has sometimes shown slightly higher scores between computer and human than between humans, presumably because human judges occasionally feel fatigue that causes variability within a single judge.

In human and AI analyses of communication, the central issue concerns the risk, stakes, or consequences that might arise from misunderstanding or ambiguity of interpretation. In many human communication situations, stakes or consequences of misunderstanding are low, so ambiguity can be tolerated. As the importance of clarity increases, humans use questions, redundancy, and dialogue to help disambiguate meaning. In many instances where AI is considered in language, the problem to be addressed is one of human resource. It may not be feasible to train highly reliable humans to conduct or analyze all the communications, or to score all student responses. However, as the consequences or risks to the participating students increase, so does the importance of reliable, precise interpretation of language or scoring.

Once again, there is value in a hybrid assessment system that compares computer-generated scores and human-generated scores. Everything is in harmony when there is a quantitative compatibility between the computer and human scores. When there is quantitative incompatibility in the human annotations, then the argument can be made that it is reasonable to substitute a computer for a human. When there is agreement among humans, but discrepancy with the computer scores, then there needs to be a discussion to resolve the incompatibilities. That would be a decision with beneficence, especially when all decisions are transparent to all parties. There is another dimension of beneficence, namely the effort of human instructors to grade essays and homework. We have never met an instructor in a large course of 300 students who enjoys grading and deeply reading thousands of essays and homework assignments.

When the first author of this article was editor of the journal *Discourse Processes,* there were many papers from multiple fields that analyzed text or conversational discourse on multiple theoretical levels and categories that attempted to capture the meaning of the discourse segments in context. Disagreements were persuasive to the extent that the judgments were abstract, opaque, complex, theoretically abstruse, and applicable to lengthier texts. Some authors tried to circumvent reporting statistics on interrater reliability by arguing that they were performing an in-depth qualitative analysis on a small number of cases to argue for or against a theoretical claim. Such arguments were never persuasive for three reasons. First, the human judges could be biased to reflect their idiosyncratic sociocultural context. Second, the selection of cases could be cherry-picked and thereby exhibit a selection bias. Third, the small sample size would not support any semblance of generalization. These problems are straightforward.

However, there are two other problems that are a bit more convoluted. Presentation of a case (here a discourse segment) that deviated from theory was once upon a time a good argument to disqualify a theory. It was an echo in an era of Popper (1958) in the philosophy of science when researchers suggested you could disqualify theories with evidence but never prove a theory as being true. The problem with this line of thinking is that we now live in a statistical world, not a binary, discrete, propositional logic, case-based world. A single case that deviates from theory is fine for rhetorical presentation, but it does not constitute empirical evidence in the statistical world. The second problem with the qualitative argumentation is that the conclusions were often unenlightening. There were many hackneyed conclusions, such as "context matters," "people from different cultures interpret language differently," and "here is my favorite new analytical system for others to use" (which rarely happens). Those qualitative studies that make a contribution tend to identify, describe, or explain new phenomena from a distinct point of view that might be masked or overlooked in quantitative studies or simply benefit from a richer treatment than is afforded in most quantitative research reports. This in turn often leads to better informed quantitative studies to codify new constructs or theoretical relationships among the constituents.

Once again, AI can be part of the solution rather than a problem. Output from computers on meaning analyses can be provided by computers, as in the case of Vasile Rus's SEMILAR systems (Rus et al., 2013) that analyzes the semantics of language segments and pairs of segments with many resources in computational linguistics from different institutions. Coh-Metrix (McNamara et al., 2014) analyzes texts on multiple levels of language and discourse, whereas Linguistic Inquiry and Word Count (Pennebaker et al., 2007) analyzes words on dozens of theory-based psychological categories. These automated approaches to language analysis in some instances mimic the coding of instances in qualitative studies, and can be compared with human judgments in a hybrid ethical approach. That is, often qualitative methods require careful training of human coders who are reliable and consistent in their judgments. Such coder characteristics are within the scope of AI capabilities and provide an efficient alternative process both for coding and examining the reliability and validity of qualitative coding systems.

There will always be ambiguity in the meaning of language. In humans this ambiguity may stem from differences in knowledge or assumptions, from lack of clarity in communicating one's intended meaning, or from intentionally obscuring meaning, perhaps to be vague in a political world or as an art form to allow for multiple interpretations in poetry or literature. To build trust in the use of AI, such artifice in communications should be minimized. Guidelines and transparency about the capabilities and accuracy of machine analysis of language should be provided, with care given to ensuring that as consequences of interpretations of language increase, so do the quality control techniques ensure that AI systems are at least no worse than human judges in interpreting meaning.

5 How Can Computer Agents Be Compared with Human Conversation Partners?

Some chapters commented on the ethics of bots and computer agents in AI systems (DeFalco & Hart; Hampton, Morrison, & Morgan; Zhang, Hu, Andrasik, & Feng). There were substantial disagreements on the quality of the agents' contributions and the need for agents to reveal their cyber-status. This is an area we have investigated in learning environments, such as AutoTutor (Graesser, 2016; Graesser et al., 2020), that help students learn by holding a conversation in natural language. Agents such as those in AutoTutor do not perfectly understand learners, but they can facilitate learning by extracting content through advances in computational linguistics, giving helpful feedback, and advancing the conversation in an adaptive rather than a rigid fashion. Human tutors also have an imperfect, approximate understanding of learner's natural language (Graesser et al., 1995; VanLehn et al., 2003), but they extract enough information to help them learn better than reading text or listening to lectures (VanLehn, 2011). Modern intelligent tutoring systems with natural language interaction facilitate learning about as well as human tutors (Graesser, 2016; VanLehn, 2011). This is an important advance because human tutors are a scarce and valuable resource.

The chapters proposed several agent designs for AI researchers to explore. There can be multiple agents in a learning environment that take on different roles (tutor, mentor, peer, game competitor) and multiple conceptual perspectives. An agent can implement different strategies with different payoff matrices, which presumably influences learning, negotiation, and collaboration. These are all promising avenues that are destined to occupy research in the future.

The authors raised some ethical issues to consider in the use of agents. One issue is whether the agent should announce it is an agent rather than a human. There are different views on this as well as tradeoffs that need to be evaluated on ethical criteria. It may not matter in some contexts. People do not routinely care whether actors in a movie believe what they say or are products of a system with scripts, directors, and other aspects of the motion picture industry. Similarly, people may not care whether a message was generated by a computer or a person as long as it is useful for the person in achieving her goals (see Hampton et al. chapter). In other contexts, however, it may matter for the agent to divulge its cyber-ness, as in the case of a stock market transaction or a medical procedure. The extent to which an agent has anthropological features that are romantically or sexually seductive may be called into question on its ethical dangers. But there are contexts when it may be appropriate, such as elderly folks who know quite well that they are physically and emotionally past their prime. Once again, simple absolute rules are not easy to formulate because there are tradeoffs, depending on context.

Wizard of Oz techniques are often implemented to have humans, rather than computer systems, generate the messages of avatars. This approach is sometimes expanded to have computers generate the message and then the human either accepting the computer message to move forward or rejecting the message and instead producing a human message. The human messages can be analyzed, augmented with AI techniques, and eventually be replaced by the automatically generated messages. The success of the system can be measured by the percentage of automatic messages that are accepted by the human wizard. At some level approaching 100% the computer is ready to replace the human. Of course, there will always be backup provisions for wild input by the user. Nevertheless, this approach illustrates the value of a hybrid system that accommodates both AI and human intelligence, with objective quantitative measures for accountability.

6 How Can Humans and AI Systems Promote Beneficence and Dignity for Humans?

Several of the chapters (D. DeFalco; J. DeFalco & Hart; Ness; Pyke, Schoenherr, & Thomson; Schoenherr & Thomson; Ulgen) emphasized that both human institutions and AI systems should protect the dignity, respect, status, and beneficence of humans. This is indeed a widely accepted assumption of the contributors of all chapters, although there is a growing trend to consider the beneficence of planet earth and its ecology of flora and fauna, with humans not uniquely as the center of the universe. One concern is that AI will replace the jobs of humans. Citizens with lower skills will be particularly affected (Autor et al., 2003; Elliott, 2017). Another concern is that there are multiple adverse consequences of AI that need to be tracked and rectified. However, the chapter contributors did not uniformly agree on the advantages and liabilities of humans versus AI systems in promoting beneficence and dignity.

Researchers in psychology, medicine, and most of the social sciences have helped identify critical issues in protecting and promoting beneficence and dignity in research with human subjects. The Belmont Report's coverage of beneficence, justice, and respect for persons is articulated in its Ethical Principles and Guidelines for the Protection of Human Subjects of Research (1978[1]). The Common Rule (NIH 45 CFR 46) is the Code of Federal Regulations that was established for the Protection of Human Subjects, which outlines the criteria and mechanisms for IRB review of human subjects research. Emanuel et al. (2000) articulated "seven critical issues" that must be scrutinized when evaluating any study on the ethical conduct of clinical research: value, scientific validity, fair subject selection, favorable risk-benefit ratio, independent review, informed consent by the participant, and respect for enrolled participants. Researchers in the medical and social sciences need to complete training on the ethical treatment of participants every two years and need to have their studies approved by a five-member Institutional Review Board. The training and bureaucratic process takes effort and time. It takes so much time that Hu and Graesser (2004) developed an adaptive intelligent tutoring system to assist high ranking officers in the Department of Defense who had to make decisions on research projects involving human subjects, but had limited time for the training of the ethical principles). This kind of ethical scrutiny addresses whether and how research is conducted. It can also be applied in the research and development of AI systems, many of which use human subjects' data in developing effective algorithms.

A sensible step for researchers in AI and other computer technologies would be to review and possibly adopt many of the principles of ethics that have evolved in the social and medical sciences for over a decade. It is important to emphasize that these are principles to guide ethical decisions, not iron-clad rules without exceptions. Whatever principles are adopted need to be transparent to both the developers of the technology and the users of the technology. Unfortunately, these ethical principles and guidelines do not address issues of developed AI systems with the general public. Government research (e.g., National Institutes of Health) and regulation agencies (e.g., Federal Drug Administration) are better analogies for the kinds of control required to protect the public. Governmental or institutional agencies have their own trade-offs, but these should be measured against the level of legitimate risks that unethical use of AI systems could pose to the public at large.

At a more local level, there are tradeoffs in many decisions on the ethical conduct of research. One noteworthy dilemma is in comparisons between an experimental treatment of a drug, for example, and a control condition without the drug. Concerns are raised that the control subjects might be deprived of the treatment and thereby not be treated fairly; this argument, of course, presupposes that the treatment is effective, where it may, in fact, be ineffective or even harmful. One way to handle this problem is to delay the administration of the treatment to the control condition for a year or so in order for them to eventually receive the treatment. Unfortunately, this solution precludes the opportunity to investigate the long-term effects of the treatment over many years or decades. Examples like these illustrate the difficulties and potential unwanted consequences of research decisions. In general, the need for continued research, not only before an AI system is released, but also during its effective lifespan in public use, are sensible protections to expect in the future.

The possibilities of tradeoffs and unwanted consequences of decisions need to be explored in AI applications. Authors of some of the chapters either asserted or presupposed that relying on AI is inherently problematic so humans need to be responsible for making the ultimate decisions. However, the authors were not uniformly on board with this position. The argument could be made that humans are imperfect (as discussed in previous sections), and that it would be impractical if not unethical to have humans make decisions at some points in the process. This dilemma motivates an analysis of tradeoffs. The AI mechanisms may be too complex to explain to most people and attempts to do so might disrupt activities of humans in ways that are threats to beneficence. At the same time, transparency of the mechanisms, the research supporting them, and the experts who are best able to evaluate the evidence should be made available to scrutiny. A relevant issue is the building of trust in systems, whether operated by humans, AI, or a hybrid approach. As humans, we typically trust in expertise (doctors, engineers) and the methods that prepare and certify that expertise. We then have at least some trust in the products of their practice—the patients they treat or the bridges they build. Similar institutional systems, as discussed in several chapters, need to be created to both regulate and build trust in the use of AI systems.

In summary, perhaps AI systems should be required to conform to a higher standard of ethics than humans for the same contexts. Effectiveness and consequential errors made by AI systems should be closely tracked and levels of precision (tolerances, like in engineering) put in place for acceptable levels of error. Extensive use of quality control including statistical probabilities, confidence intervals, and effect sizes should be monitored and reported. Such a system of continuous monitoring will both build confidence in systems and help enhance systems for future use.

Similarly, the humans behind the development and implementation of AI systems need to follow both ethical guidelines and associated methods to build trust in the AI. Transparency should be accessible in important steps of the process: the methods used to create the AI system; the methods for human monitoring /safeguarding /cross-checking the system; the evaluation of the level of consequences and risk of different types of errors. System-wide end users need not be forced to review this information, but they should be made aware of its availability, in plain language whenever feasible. Most end users would not likely access such information, but its availability to the public might help developers as well as users when questions are raised. End users are not likely to scrutinize fine-grain details of a complex process, such as decisions made at particular branch points in an educational game with stealth adaptive learning (Shute, 2015). Transparency also becomes more debatable in open learning environments (Bull & Kay, 2007) that allow students to inspect their learning and psychological attributes. Some knowledge, skills, and abilities would make sense to show to a student, but not others, such as a low creativity score on a subject matter the student is passionate about. An IRB decision would need to evaluate such tradeoffs on matters of transparency at different levels of abstraction.

Ideally, humans would convene to agree on principles and guidelines to allow AI to augment decision making in areas that humans are known to be weak: formal logic, statistical reasoning, historic biases, and group think. When there is process and outcome-based evaluation as well as direct comparisons of human versus AI decisions/outcomes, there is some hope for improving the integrity of the AI systems toward beneficence and respect for humans. Importantly, it also provides an opportunity to reflect on human aspirations toward promoting social beneficence and respect. We may find that the methods we have designed to promote ethical research, development, and use of our science, including AI systems, might also have value applied in other human activities, institutions, and endeavors.

7 Conclusion

In closing, this chapter (and by extension, book) hopefully leaves the reader with a few simple messages. Humans are far from perfect in making ethical decisions. There are nearly always tradeoffs in ethical decisions. It is worthwhile to consider hybrid systems that compare output of human and AI systems, as in the case of computer agents and analyses of meaning. AI can be part of the solution to resolving ethical dilemmas. This includes application to hypothetical cases that may or may not have already occurred in the real world, and even to ethical dilemmas brought about by the very presence of AI. To guide this process, AI can benefit from decades of policies in the social and medical sciences on the ethical treatment of humans in research.

Note

1 See https://www.hhs.gov/ohrp/regulations-and-policy/belmont-report/index.html.

References

Ahad, M. A., Tripathi, G., & Agarwal, P. (2018). Learning analytics for IoE based educational model using deep learning techniques: Architecture, challenges and applications. *Smart Learning Environments*, 5(1), 1–16.

American Educational Research Association, American Psychological Association, & National Council on Measurement in Education. (1999). *Standards for educational and psychological testing*. Joint Committee on Standards for Education and Psychological Testing.

Autor, D., Levy, F., & Murnane, R. J. (2003). The skill content of recent technological change: An empirical exploration. *Quarterly Journal of Economics*, 118, 1279–1334.

Baker, R. S. (2020). *Big data and education* (6th ed.). Philadelphia, PA: University of Pennsylvania.

Bull, S., & Kay, J. (2007). Student models that invite the learner in: The SMILI: ()Open learner modelling framework. *International Journal of Artificial Intelligence in Education*, 17(2), 89–120.

Cai, Z., Graesser, A. C., Forsyth, C., Burkett, C., Millis, K., Wallace, P., Halpern, D., & Butler, H. (2011). Trialog in ARIES: User input assessment in an intelligent tutoring system. In W. Chen & S. Li (Eds.), *Proceedings of the 3rd IEEE International Conference on Intelligent Computing and Intelligent Systems* (pp. 429–433). Guangzhou: IEEE Press.

Campbell, D. T., & Cook, T. D. (1979). *Quasi-experimentation*. Chicago, IL: Rand Mc-Nally.

Chen, S, Fang, Y., Shi, G., Sabatini, J., Greenberg, D., Frijters, J., & Graesser, A.C. (2021). Automated disengagement tracking within an intelligent tutoring system. *Frontiers in Artificial Intelligence*, 3,1–16.

D'Mello, S. K., & Graesser, A. C. (2012). AutoTutor and affective AutoTutor: Learning by talking with cognitively and emotionally intelligent computers that talk back. *ACM Transactions on Interactive Intelligent Systems*, 2(4), 23:2–23:29.

Elliott, S. (2017). *Computers and the future of skill demand*. Paris: OECD.

Emanuel, E. J., Wendler, D., & Grady, C. (2000). What makes clinical research ethical? *Journal of the American Medical Association*, 283, 2701–2711.

Fiske, S. T. (1998). Stereotyping, prejudice, and discrimination. In D. T. Gilbert, S. T. Fiske, & G. Lindzey (Eds.), *Handbook of social psychology* (4th ed., Vol. 2, pp. 357–411). New York: McGraw-Hill.

Franklin, S. (1995). *Artificial minds*. Cambridge, MA: MIT Press.

Gersten, R., & Hitchcock, J. (2009). What is credible evidence in education? The role of the what works Clearinghouse in informing the process. *What Counts as Credible Evidence in Applied Research and Evaluation Practice*, 78–95.

Gladwell, M. (2019). *Talking to strangers: What we should know about the people we don't know.* New York: Little Brown and Company.

Graesser, A. C. (1995). Imagine law without simple rules. *Contemporary Psychology, 40,* 143–144.

Graesser, A. C. (2016). Conversations with AutoTutor help students learn. *International Journal of Artificial Intelligence in Education, 26*(1), 124–132.

Graesser, A. C., Foltz, P. W., Rosen, Y., Shaffer, D. W., Forsyth, C., & Germany, M. L. (2018). Challenges of assessing collaborative problem solving. E. Care et al. (eds.), In *Assessment and teaching of 21st century skills* (pp. 75–91). Cham: Springer.

Graesser, A. C., Hu, X., Rus, V., & Cai, Z. (2020). Conversation-based learning and assessment environments. In D. Yan, A. Rupp, & P. Foltz (Eds.), *Handbook of automated scoring: Theory into practice* (pp. 383–402). New York: CRC Press/Taylor and Francis.

Graesser, A. C., Person, N. K., & Magliano, J. P. (1995). Collaborative dialogue patterns in naturalistic one-to-one tutoring. *Applied cognitive psychology, 9*(6), 495–522.

Graesser, A. C., Wiemer-Hastings, K., Kreuz, R., Wiemer-Hastings, P., & Marquis, K. (2000). QUAID: A questionnaire evaluation aid for survey methodologists. *Behavior Research Methods, Instruments, & Computers, 32*(2), 254–262.

Hu, X., & Graesser, A. C. (2004). Human Use Regulatory Affairs Advisor (HURAA): Learning about research ethics with intelligent learning modules. *Behavior Research Methods, Instruments, and Computers, 36,* 241–249.

Janis, I. L. (1982). *Groupthink.* Boston, MA: Houghton Mifflin.

Johnson-Laird, P. N. (2010). Mental models and human reasoning. *Proceedings of the National Academy of Sciences of the United States of America, 107,* 18243–18250.

Kahneman, D. (2011). *Thinking, fast and slow.* New York: Farrar, Straus and Giroux.

Kahneman, D., Sibony, O., & Sunstein, C. S. (2021). *Noise: A flaw in human judgment.* New York: Little Brown.

Kahneman, D., Slovic, P., & Tversky, A. (1982). *Judgment under uncertainty: Heuristics and biases.* New York: Cambridge University Press.

McNamara, D. S., Graesser, A. C., McCarthy, P. M., & Cai, Z. (2014). *Automated evaluation of text and discourse with Coh-Metrix.* New York: Cambridge University Press.

Messick, S. (1995). Standards of validity and the validity of standards in performance assessment. *Educational Measurement: Issues and Practice, 14*(4), 5–8.

Newell, A., & Simon, H. A. (1972). *Human problem solving* (Vol. 104, No. 9). Englewood Cliffs, NJ: Prentice-hall.

Pennebaker, J. W., Booth, R. J., & Francis, M. E. (2007). *Linguistic inquiry and word count: LIWC* [Computer software]. Austin, TX: liwc. net, p. 135.

Petty, R. E., & Cacioppo, J. T. (1986) The elaboration likelihood model of persuasion. *Advances in Experimental Social Psychology, 19,* 123–205.

Pinker, S. (2018). *Enlightenment now: The case for reason, science, humanism, and progress*. New York: Penguin Random House.
Popper, K.(1958). *The logic of scientific discovery*. New York: Routledge.
Rapp, D., & Braasch, J. (2015) (Eds.). *Processing inaccurate information: Theoretical and applied perspectives from cognitive science and the educational sciences*. Cambridge, MA: MIT Press.
Reynolds, C. R., Altmann, R. A., & Allen, D. N. (2021). The problem of bias in psychological assessment. (Eds., Reynolds, a., Altmann, R., & Allen, D.) In *Mastering modern psychological testing* (pp. 573–613). Cham: Springer.
Rips, L. J. (1994). *The psychology of proof: Deduction in human thinking*. Cambridge, MA: MIT Press.
Rus, V., Lintean, M., Banjade, R., Niraula, N. B., & Stefanescu, D. (2013). SEMILAR: The semantic similarity toolkit. In *Proceedings of the 51st annual meeting of the association for computational linguistics: System demonstrations* (pp. 163–168).
Shute, V. J. (2015). Stealth assessment. Stealth assessment. In J. M. Spector (Ed.), *Encyclopedia of educational technology* (pp. 675–678). Thousand Oaks, CA: Sage Publications.
Silverman, A. E. (1993). *Mind, machine, and metaphor: An essay on artificial-intelligence and legal reasoning*. New York: Routledge.
United States. National Commission for the Protection of Human Subjects of Biomedical, & Behavioral Research. (1978). *The Belmont report: Ethical principles and guidelines for the protection of human subjects of research* (Vol. 2). The Commission.
Van Lehn, K. (2011). The relative effectiveness of human tutoring, intelligent tutoring systems and other tutoring systems. *Educational Psychologist*, 46(4), 197–221.
Van Lehn, K., Siler, S., Murray, C., Yamauchi, T., & Baggett, W. B. (2003). Why do only some events cause learning during human tutoring? *Cognition and Instruction*, 21(3), 209–249.
Yan, D., Rupp, A. A., & Foltz, P. W. (Eds.). (2020) *Handbook of automated scoring: Theory into practice*. New York: CRC Press/Taylor and Francis.

Index

Note: **Bold** page numbers refer to tables; *italic* page numbers refer to figures and page numbers followed by "n" denote endnotes.

accountability 12, 41, 192
actor bias 52
acute stress 131
Ada and Grace, virtual twins 127
adaptive instructional systems (AIS) 155n1
Affective AutoTutor 85–86
AI-based decision support system 25
AI-driven systems 135
Amazon 169, 173
Amazon AI system 177
analogical reasoning 142, 154
analogy-based frameworks 151
anomaly-based botnet detection 153
anomaly detection 152–154
anthropomorphism 86
APLUS 91–92
Aristotle: *Nicomachean Ethics* 129
Arquilla, J. 143, 144
artificial agents 112, 114
artificial intelligence (AI) 1, 45, 185; behavior of organic and 116n2; cultural awareness for 50; ethics 2 (*see also* ethics); human resources use case for 162–164; innovation 10; and racial bias 40
artificial intelligence (AI) bias 45; actor bias 52; case selection bias 51–52; levels of analysis 51–54; media bias 52–53; process bias 53; semantic bias 53–54; temporal bias 53
artificial intelligence (AI) developers and design, bias in 167–173

Artificial Intelligence Ethics Framework for the Intelligence Community 43
artificial intelligence (AI) selection systems: bias in AI developers and design 167–173; bias in interpretation and use of AI recommendations 178–179; bias in training data 173–178; potential sources of bias in 166–167
artificial intelligence (AI) systems: advantages of 165; ethics of 185; humans *vs.* (*see* humans *vs.* AI systems); meta-level of analysis 164; promote beneficence and dignity for humans 196–199; symbolic and statistical algorithms in 187
artificially intelligent agents 102, 103
ASSISTments tool 88–89
Atkinson, R. 68
attention deficit hyperactivity disorder (ADHD) 114, 115
Augustine of Hippo (354–430 CE) 14
Authorable Virtual Peer 127
authoring tool 91
autism spectrum disorder (ASD) 114, 115
automated essay grading 193
automated facial recognition machine learning algorithms 40
autonomous and intelligent systems (A/IS) 141, 142, 145, 152
autonomous weapons, use of 22, 23

204   Index

AutoTutor 89–90, 195
AXIS 89

Balanced Inventory of Desirable
  Responding 73
Bateson, G. 67
Batey, M. 132
Berkeley Center for Teaching and
  Learning 78, 78–79
Betty's Brain 91
Betz, D. J. 144
bias: in AI developers and design
  167–173; in AI selection systems
  166–167; in humans vs. AI systems
  190–192; in interpretation and use
  of AI recommendations 178–179; in
  training data 173–178
big data 52
Big Five 132, 133
Bilateral Negotiation (BiLAT)
  simulation 128
Bill of Rights, democratic value of 17
Biocca, F. 86
Bloom, B. 108
"Blue Box" 127–128
Blue Match system 164
Bostrom, N. 2
Brainerd, C. J. 67
Brief Fear of Negative Evaluation
  Scale 73
Buolamwini, J. 40
Burrell, Lolita 3

CAMEO 48, 50, 54
CAMEO codes 50, 54
Campbell, D. T. 191
capacity, for rational conduct 15
Captain, S. 163
case selection bias 51–52
Cassell, J. 127
Chicken Dilemma 150
China, responsibility 12
Christian theology 14
Christian thinking 14
Clarke, Arthur C.: *The Pacifist* 125
class clown agent 110, 111
classmate behaviors, ITSs 90
classroom: construct of mind and
  intersubjectivity implications 66–71,
  67; ITSs implement classmate
  behaviors 90; ITSs implement
  learner behaviors 91–92; ITSs

implement teacher behaviors 89–90;
  technology implications in 72–77;
  technology in 64
Code of Federal Regulations 197
cognitive load, pedagogical agent
  87–88
cognitive theory 64; of learning 104;
  pragmatic assessment of 78, 78–79
Coh-Metrix 194
Cold War 53
Collins, B. 134
Common Rule 197
communication 71; cross-cultural
  50; human and AI analyses of 193;
  interpersonal 114; between learners
  and conversational ITSs 85; parent–
  teacher 96
communication disorders 114, 115
communicative systems 65
compositional fairness 168
"computational propaganda" 23
computer agents vs. human
  conversation partners 195–196
computer decision-making systems,
  based on machine learning
  algorithms 167
Computer Expression Recognition
  Toolbox 73
conflict 46–47, 53; Google N-gram
  for war and 46, 47; perspectives
  variations on political events 48, 48
conflict-cooperation continuum 49
Confucianism human dignity
  functions 14
consciousness 75
construct of mind 66–71, 67
contemporary invisible risks: AI
  and racial bias 40; identifying
  individuals using DNA 38–39;
  private traits using digital records 39
contemporary language 53
context dependent salience
  assignment 70
continual cultural process 65
conventional student–teacher
  ratios 103
conversational agents 105–107
conversational interaction 103
conversational ITS 109, 111, 115
conversation-based instruction 3
conversation-based intelligent tutoring
  systems 84

Cook, T. D. 191
Counter Insurgency and Stability Operations 128
Covariation Model of Attribution 151
COVID-19 pandemic, in UK and Europe 10
cross-cultural communication 50
CSILE 105
cultural biases 49
culture: conveyors of 48–51, 50; and language 45, 53; through kaleidoscope 54–56, 55
Currier, J. M. 130
"cyber cold war" 145
cyber ethics 141
cyberoperations 143–145; analogies for 146, **147–148**
cyberpower 143
cybersecurity 142, 146
cyberspace 143
cyberwar 143–145, 154
cyberwarfare techniques 141

Dangwal, R. 76, 77
data protection rights 22
data science 34; risks of 43; use of 35
Davis, F. D. 92
decision-making process 3–4, 131–132, 145; in AI systems 27; moral judgment 133–134; moral sensitivity 134–135; personality traits 132–133
decision-support 3
deep-level reasoning questions 108
deep neural networks 49
DeFalco, Jeanine A. 1, 3
Defining Issues Test (DIT) 133–134
"Degree of Mindfulness" 134
description-experience gap 149
de-skilling 13
Dewey, J. 42
dialogue-based approaches 106
dialogue-based intelligent tutoring systems 106
differential item analysis 191
digital records, private traits using 39
digital technologies, use of 13
digital technology acceptance 92; learner's digital technology acceptance 94–95, 95; parent's digital technology acceptance 95–96; TAM roles 92–93, 93;
teacher's digital technology acceptance 93–94
dignitas 14
discourse-based systems 105
discourse learning 103–105
Discourse on Language (Foucault) 41
distance learning 102
divergent ethical/legal initiatives 12
DNA, identifying individuals using 38–39
domain model, intelligent tutoring systems 85
Doty, J. 134, 135
duty, human dignity as 14–17
Duval, T. S. 75
dyslexia 114–115

Ebbinghaus-like memory drum tasks 69
Edison, Thomas 63
education 44; digital technology acceptance in 92–96; purpose of 72; and training 41–42
Efthimion, P. G. 153
ELIZA 125
Elvira, M. 176
Emanuel, E. J. 197
embodied conversational agent 127
Emergent Leader Immersive Training Environment (ELITE) 128
emotional regulation, technology on 73, 74
English language 48, 51
epistemic vigilance 105
"epistemological understanding" 105
equality 17
Erlich, Y. 38
ethical affordance 142
ethical dilemmas 130–131
Ethical Principles and Guidelines for the Protection of Human Subjects of Research 197
Ethical Sensitivity Mindfulness Model (ESMM) 134
ethics: and decision making 131–135; history of 185; and military 129–130; training and certification in 186
ethics education 129, 131
EU General Data Protection Regulation (GDPR) 10; and Convention 108+ 22, 25, 27

European Molecular Biology Laboratory (EMBL) 37
Europe, COVID-19 pandemic in 10
event codes 49, 52, 53
event data 46, 50; cultural and linguistic bias in 55; de-biasing for social science research 54; encoding of 53, 54; generation of 45, 46, 48–51; identifying patterns in 48
evocative analogies 154
exclusion, systems of 41
Explainable AI (XAI) 169, 171

facial detection algorithms 176
Fairness Flow 178
fairness metrics 172, 173, 180
FATE framework 172
Fitts' Index of Performance 68
five-step tutoring frame 90
folly 41
Foucault, M.: *Discourse on Language* 41
fourth revolution 11, 26
freeing/enslaving 12–13
Frey, David 3
Furnham, A. 132
fused interaction 111–112

game analogy 145–146, *146*, **147–148**, 149
game theoretic models 145
game theory 145–146, *146*, **147–148**, 149
Gans, R. 169
Gates, Bill 2
GDELT 52
GEDmatch 38
Gee, M. 164, 173
gender 94
Geneva Conventions (1949): Articles 1(2) of 23; Articles 75 of 23; Commom Article 3 of 23, 29n12
German Basic Law, Article 1(1) of 16
German constitutional law 16, 23
germane cognitive load 86
Germany 35–37
Gibson, W. 143
Godkin, L. 134
Goldberg, E. 69, 70
Golden State Killer 38
Goldstein coding scheme 55
good student agent 110
Google 165, 169

Google N-gram 46, *47*
Graepel, T. 39
Graesser, A. C. 4, 108, 197
Granovetter, M. S. 153
Great Library of Alexandria 63
group fairness 172
Group Scribbles 105

hackers, techno-bureaucracy of 11–12
Hampton, Andrew 1
Han, Byung-Chul 11
Hart, John 3
"HeLa" cells 37
Henrietta Lacks 37–38
Heuss, Theodor 16
HEXACO 132–133
HEXACO-PI scale 132; advantages of 133
HEXACO–60 scales 133
Hinduism 14
HireVue system 163
Hitler, Adolf 35, 36
Hofer, M. 65
"Hole in the Wall" project 76
Holland, J. M. 130
Hollerith tabulation machines 35, 36
Honesty–Humility trait 132
Hsu, Y.-Y. 96
human agents: with autonomy 18–19; with rational capacity to exercise reasoning, judgement, and choice 19–21; respectful prioritization of human needs 26; respectful treatment of 21–27; respecting AI limitations 25; respecting rights 22–25
human beings status, recognition of: agents with autonomy 18–19; agents with rational capacity to exercise reasoning, judgement, and choice 19–21
human-centric approach 3, 22, 26–28
human conversation partners *vs.* computer agents 195–196
human dignity 27; AI innovation and impact on 10–13; detrimental impact of AI on 9–10; as fundamental value and right 17; hierarchy of 22; in international legal instruments and constitutions 15–17; moral value of 10; as status and respectful treatment 17–26; as universal moral value, right, and duty 14–17

"human-in-the-loop" decision making 178
human knowledge acquisition 73
human learners 90, 91, 96, 107–110, 112
human organizational developers 168–169
human resources (HR) 162–163; AI systems use in 163–164; benefits in AI system 164–166
humans 74; *vs.* computer framed interaction 73, 74; development of knowledge 70–71; relationship between intelligent anthropomorphic entities and 86; tradition for diffusing ideas 66
humans *vs.* AI systems: forms of bias exist in 190–192; indeterminacy of meaning among 192–195; rational reasoning 186–190
human-to-human social interchange 73
human virtual agents 125–126, 135
Hu, X. 197

IBM: Blue Match system 164; retrospective invisible risks 35–37
ICCPR *see* International Covenant on Civil and Political Rights (ICCPR)
ICEWS 52
Ifenthaler, D. 94
Im, I. 94
information domains, analogies of 142–143; cyberoperations and cyberwar 143–145; interactional schemata to social agent development and detection 151–154; *kriegsspiel* 145–146, *146*, **147–148**, 149; schematic models of relational structures 149–151, **150**
information technology 63
Information Theory 68
inherent dignity 16, 17
institutional power structures 41–43
Institutional Review Board (IRB) 185, 186, 197
instructor agents 111
Intelligence Advanced Research Projects Activity 40
intelligent agents 84; anthropomorphic features of 86
intelligent tutoring systems (ITSs) 84, 105, 114; implement classmate behaviors 90; implement learner behaviors 91–92; implement teacher behaviors 89–90; pedagogical agents in 85–88; as software tools 88–89
interactional schemata, to social agent development and detection 151–154
Interdependence Theory 149–153
intermediaries 75–76
intermittently artificial instructors 113–114
internally persuasive discourse 103
international agents 45
International Baccalaureate Organisation 10
International Convention of Psychological Sciences (ICPS) 2
International Covenant on Civil and Political Rights (ICCPR) 16; Article 10 of 16
International Covenant on Economic, Social and Cultural Rights (ICESCR), Article 13 of 16
Internet 63, 64
intersubjectivity 71; implications of 66–71, **67**
Interviewed system 163
interviews 179
invisible risks: contemporary 38–40; retrospective 34–38
Ishowo-Oloko, F. 152
Islam 14
ITSs *see* intelligent tutoring systems (ITSs)

James, William 64
Joint Declaration on Freedom of Expression and Elections in the Digital Age 23, 27–28, 29n13

kaleidoscope process, language and culture through 54–56, *55*
Kant, I. 79
Kantian deontological ethics 26
Kantian human dignity 14–15
Kant, Immanuel 14, 15; "disgraceful punishments" 23; secular theory 17–18
Katz, J. H. 169
Kelley, H. H. 151, 154
Kiesler, S. 145
Kiss, Bow, or Shake Hands (Morrison and Conaway) 46

Kleinberg, J. 163
knowledge: components 102; social construction of 104
Koru system 166
Kosinski, M. 39
kriegsspiel 145–146, *146,* **147–148,** 149
Kruger, A. 104

language: culture and 45, 53; Google N-gram for "war" and "conflict" by 46, 47; indeterministic nature of 64–66; relationships between thought and 71–72, **72**; SVO (subject verb object) 53; through kaleidoscope 54–56, *55*; of United Nations 51
Latent Semantic Analysis 49
learner digital technology acceptance 94–95, *95*
learner model, intelligent tutoring systems 85
learners 85, 86; behaviors, ITSs 91–92; next generation of 102
learning: compare and contrast of Vygotsky and Piaget 71, **72**; definition of 78, **78–79**; quality of 108; self-organized 76; social activity 87
learning-by-teaching method 91–92
learning management systems 102
legal liability 12
legal reasoning 189
legal responsibility 12
Lewin, K. 149
liability 12; responsibility and 35
Life Events instrument 134–135
Linguistic Inquiry 194
logical reasoning 187–188
Lucas, G. 73, 74

machine learning (ML) 45, 49; cultural awareness for 50
machine learning algorithms 170; computer decision-making systems based on 167
MacIntyre, A. 134, 135
Malott, J. 130
Manyika, J. 164
Marlowe, C. M. 176
May, E. 143
mechanistic derived models 68

mechanistic organismic world views, presuppositions of 66, **67**
media bias 52–53
memory 69; variability in activity of 71
mental health 130–131
Mental Health Assessment Team (MHAT) survey 129
Messick, S. 191
MIEs *see* morally injurious experiences (MIEs)
military: ethical decision-making capabilities of 125; ethics and 129–130; simulators 127–128; in training language and cultural skills 127
Miller, F. A. 169
mind: construct of 66–71, **67**; definition of 68; organizing structures of 70–71
Mitra, S. 76, 77
modern AI systems 189–190
Moore's Law 34
moral injury 130–131
morality 42
moral judgment 133–134
morally injurious experiences (MIEs) 130, 131
moral sensitivity 134–135
motivation: issue for ADHD learners 115; pedagogical agent 87
multiple virtual agents 109–111, *110*
mundane plausible reasoning 187
Mundy, L. 168
Musk, Elon 2

Naïve Bayes classifier 90
Nami, F. 95
Narayanan, A. 38, 39
Nash, J. F. 145, 149
National Institute of Standards and Technology (NIST) 40
natural language processing (NLP) techniques 46, 106
Nazi Germany 36, 37
Ness, James 3
Ness, J. W. 65
neural network (NN) 189
Neustadt, R. 143
Nicomachean Ethics (Aristotle) 129
Nobel, Emil Oskar 43
"non-interpreted thesis" 16
non-state actors 24, 28

Norwegian Data Protection
  Authority 10
Nowak, K. L. 86

Objective Self 75
obstacle respect 26
occasionally artificial peers 112–113
one-to-one (human) tutoring 105–106
open-ended approach 106
open-ended learning environment 91
operationalizing culture 51
organismic world views,
  presuppositions of 66, **67**
organizational change 71
Organization for Security and
  Co-operation in Europe
  Representative on Freedom of the
  Media 29n13
Organization of American States
  Special Rapporteur on Freedom of
  Expression 29n13
oversampling 175

The Pacifist (Clarke) 125
parent–child communication 96
parent digital technology acceptance
  95–96
parent–teacher communication 96
pay-off matrices 150, 151
pedagogical agents 85;
  anthropomorphism 86; effects of
  87–88; ethically responsible usage
  of 97; in ITSs 85–86
pedagogical approaches 111;
  flexibility in 109
pedagogical intelligent agents 84
pedagogical model, intelligent tutoring
  systems 85
peer-to-peer conversation 111
Peirce, C. S. 65, 66
permissions 41
personality 74
personality traits 132–133
Petrarch-2 52
Piaget, J. 71, **72**, 76
Pinker, S. 188
Popper, K. R. 66, 194
pragmatic assessment, of cognitive
  theory 78, **78–79**
predicted performance: candidate
  properties used to 177–178;
  measures of 176–177

presence, pedagogical agent 87
pre-trial bail risk algorithms 10–11
Pretrial Justice Institute 11
Pribram, K. 70
Prisoner's Dilemma 146, 149, 154
privacy rights 22
process bias 53
Protection of Human Subjects 197
psychology 2, 3, 24, 185, 188,
  190, 197
*Psychology, Ethics, and Artificial
  Intelligence* 2
psychopolitics 11
Pyke, Aryn 4

quality assurance metrics 172–173

racial bias 40
rational conduct, capacity for 15
Rattray, G. J. 144
reasonable effort 24
reconstruction problem 177
recruiting systems 178
regulators, techno-bureaucracy of
  11–12
relational structures, schematic models
  of 149–151, **150**
representative sampling 191
responsibility 12; and liability 35
Rest, J. R. 134
retrospective invisible risks 34–35;
  Henrietta Lacks 37–38; IBM 35–37
right, human dignity as 14–17
robots 77
Rogers-Sirin, L. 134
Ronfeldt, D. 143, 144
RoundPegg system 163
Russian-Ukrainian conflict 50, **50**
Rus, Vasile 194

Sabitini, John 4
sadness: degree of 73; feelings of 74
Savage-Rumbaugh, S. 104
schema-based framework 154; for
  social interaction 151
schema-based models 154
Schoenherr, F. Jordan Richard 4, 150
Schrodt, P. A. 46
Schweinbenz, V. 94
screening decisions 163, 164
"security through obscurity" 38
self-awareness 75, 76

self, crisis of: freeing/enslaving 12–13; techno-bureaucracy of hackers and regulators 11–12
self-disclosure, degree of 73
self-efficacy 95
self-learning algorithms 12
self-organized learning 76
semantic bias 53–54
SEMILAR systems 194
semiotic 65, 66, 68
sensemaking 131
Shannon, C. E. 68
Sharable Knowledge Online (SKO) 109–110
Shiffrin, R. 68
Shmatikov, V. 39
signal detection theory 174
signaling patterns 69
signature-based botnet detection 153
Silberg, J. 164
Silverman, Alexander 189
SimStudent 91–92
simulations 128
Sirin, S. 134
Skinner, B. F. 69
social agents: development and detection 151–152; social bots and anomaly detection 152–154; trust and exchange with virtual agents 152
social bots 152–154
social cognitive processes 141
social dilemmas 146, *146*, 150; payoff structure and types of 150, **150**
social interaction: pedagogical agent 88; schema-based framework for 151
social media: detecting bots on 153; user interaction within 153
sociocultural theories of learning 104
"socio-epistemological engineering" 105
socio-epistemological mechanism 105
South African Constitution 17
South African Constitutional Court 17
Sproull, L. 145
Stackelberg Competitions 145–146
statistical reasoning 188
Stevens, T. 144
Stiglitz, Joseph 9
Stillwell, D. 39
stress, on combat readiness and recovery 130–131

structured interviews 179
struggling learner 109, 111
struggling student agent 110
St. Thomas Aquinas 14
Subjective Self 75
supervised discourse learning 103
SVO (subject verb object) language 53

Tactical Language Training 128
Tactical Questioning simulation 128
TalentBin 163, 178
TAM *see* Technology Acceptance Model (TAM)
Tambe, P. 162, 163
Tartaro, A. 127
teacher behaviors, ITSs 89–90
teacher digital technology acceptance 93–94
teacher–learner dynamic of dialogues 107
teacher–peer exchanges 107
techno-bureaucracy, of hackers and regulators 11–12
technology 66; in classroom 64; on emotional regulation 73, 74; implications in classroom 72–77; unequivocal beneficence of 76; wealth of information 77
Technology Acceptance Model (TAM) 92–93, *93*; learner digital technology acceptance 94–95, *95*; parent digital technology acceptance 95–96; teacher digital technology acceptance 93–94
technology-based solutions 13
temporal bias 53
"temporal dominant foci" 70
Teo, T. 94
theories of cognitive apprenticeship 104
theory of learning-by-teaching 91
Thomson, R. 4, 150
thought, relationships between language and 71–72, **72**
three-stage candidate selection process 166
Thurstone method 65
Tiku, N. 168
Tomasello, M. 104
Town, R. 176
training data, bias in 173; candidate properties used to predict performance 177–178; measures of

predicted performance 176–177; types of potentially good performers 173–176
transparency 198–199
Tressell, Robert 9
trialogues 107–109
Tsuei, M. 96
Turing, A. M. 125
Turing Test 115
tutor agent 110

ubuntu 14
UDHR *see* Universal Declaration of Human Rights (UDHR)
uinsupervised word meaning representations 49
UK: COVID-19 pandemic in 10; legal responsibility, accountability, and legal liability 12
unduly hinder access 24
Unified Theory of Acceptance and Use of Technology (UTAUT) 92–93, 93; learner digital technology acceptance 94–95, 95; parent digital technology acceptance 95–96; teacher digital technology acceptance 93–94
United Nations, official languages of 51
United Nations Special Rapporteur on Freedom of Opinion and Expression 29n13
United States, pre-trial bail risk algorithms 11
Universal Declaration of Human Rights (UDHR) 15; Article 1 of 16; Article 22 of 16; Article 23(3) of 16
universal moral value, human dignity as 14–17
unstructured interviews 179
user interfaces, intelligent tutoring systems 85

UTAUT *see* Unified Theory of Acceptance and Use of Technology (UTAUT)

Vaezi, S. 95
Valentine, S. 134
Van Kleek, M. 26
Varol, O. 154
VIBRANT 105
virtual agents 88, 90, 91; fused interaction 111–112; multiple 109–111, *110*; surrogate in army 127–129; trust and exchange with 152
Virtual Classmates 90
virtual classroom 108
virtual human 125, 128, 135; history 125–127, **126**
virtual tutor 90
Vygotsky, L. 71, **72**, 76, 104

Waddington, C. H. 71
Waldron, J. 15
Warner, C. H. 130
Wasdyke, C. 65
Watson, Thomas 35, 36, 126
wealth of information 63, 64, 77
Weizenbaum, J. 125
Western/American bias 49
Whitehead, A. N. 66
Wicklund, R. A. 75
Wittgenstein, L. 64, 69
Wizard of Oz techniques 113–114, 196
Word Count 194
word embedding 48–51, **50**, 52, 54
word meanings 49
word2vec 49
World Values Survey Cultural Map 51

Xu, D. 134, 135

Zhou, M. 94
Zhu, S. 95

For Product Safety Concerns and Information please contact our EU representative GPSR@taylorandfrancis.com
Taylor & Francis Verlag GmbH, Kaufingerstraße 24, 80331 München, Germany